THE NATURE OF THERMODYNAMICS

LONDON : HUMPHREY MILFORD
OXFORD UNIVERSITY PRESS

THE NATURE OF THERMODYNAMICS

BY

P. W. BRIDGMAN

HOLLIS PROFESSOR OF MATHEMATICS AND
NATURAL PHILOSOPHY IN HARVARD UNIVERSITY

CAMBRIDGE · MASSACHUSETTS
HARVARD UNIVERSITY PRESS
1941

PRINTED AT THE HARVARD UNIVERSITY PRESS
CAMBRIDGE, MASSACHUSETTS, U.S.A.

CONTENTS

INTRODUCTION

NO ANALYSIS is self-terminating, but it can always be pushed indefinitely with continually accumulating refinements. The extent to which I push my analysis of a physical situation will be determined by my tastes, interests, and purposes. Tastes are no less varied among people who enjoy analysis than among those whose activities are less articulated; there are, however, two outstandingly different sorts of temperament which are determinative with respect to the broad sort of analysis that is produced. There is the precise meticulous temperament which insists on an analysis of the greatest possible precision. A method of ensuring this, which has attained a considerable popularity of late because the desired precision is attained almost automatically, is to throw the analysis into postulational form, deducing consequences from the postulates by the operations of pure logic. Such an analysis serves certain definite purposes; in particular one may feel a security in the conclusions of such an analysis which is impossible in a less formal structure. But this I suspect is not the whole reason why those whose analysis runs to this form actually do it; it must be that they enjoy such rigorously precise activities for their own sake. Personally, my tastes place me in that other group which does not take an intrinsic pleasure in the elaboration of a logically flawless and complete structure. I would undergo the labor of constructing such a structure only as it might be necessary for some more compelling purpose.

My general purpose in attempting any analysis in physics is the attainment of "understanding." What understanding is, is an individual matter, which changes with time; I can see that my own criteria of understanding have developed with the practice of my critical activities. It may be that the need for understanding is adequately met for some people by a reduction to a logical formalism, but for me so many problems of understanding obtrude themselves among the operations of logic itself that the gain in understanding which I might anticipate from an analysis poured into a logically rigorous mold is so slight as not to provide sufficient incentive to the labor of an analysis which my tastes find arid.

All of this is by way of apology (or rationalization) because this book is not going to develop along the lines which some people perhaps feel that it should. Continual use since the publication of my *Logic of Modern Physics* in 1926 has convinced me of the undoubted utility of the "operational" method of analysis in physics. At the same time other people have criticized the indefiniteness of the fundamental notions. I recognize that it would certainly be profitable to attempt to get more precision into the fundamental notion of "operation" itself, and to push the analysis further in certain directions, as for example, in trying to formulate a set of simple operations for some consistent scheme of physics, or in trying to find how much latitude is feasible in such a set. These are undoubtedly questions of interest and importance, and perhaps I shall attempt a restricted discussion of some aspects of them later, but I feel that there are other matters which are more pressing than the development of such a structure. These other considerations have presented themselves in the course of my continued use of the operational method since the publication of my *Logic*.

Perhaps most important is a growing appreciation of the great complication of the operations actually used by physicists. It seems impossible to analyze these operations sharply into different elements, but nevertheless there are rough qualitative features which can be distinguished by anyone with experience, and which I believe it is profitable to recognize. Thus there are certain operations which are predominantly instrumental, as in measuring the length of an object by successive applications of a meter stick, but even here there is an inseparable "mental" element, that of counting the number of applications. That these "mental" operations are often an essential part of the instrumental operations is obvious when one considers any measurement that involves "derived" quantities. A viscosity, for example, involves a "mental" analysis into the quantities that entered the definition of viscosity. Theoretical activities of modern physicists involve a continually increasing mental component; theoretical physicists cannot get along without mathematics or without the use of many constructions which are predominantly mathematical in character, as witness the *psi* function of wave mechanics. Since the visible work of mathematicians can be adequately described in terms of the manipulation of written symbols, a suggestive characterization of mathematical operations is "paper and pencil" operations. The mere form of this characterization shows the futility of seeking categorical sharpness in this field, for are not paper and pencil "instruments"? and is not any "mental" operation of mathematics thereby reduced to an instrumental operation? Or for that matter is not a hired computer an instrument, particularly if the computer is shut up in a box unknown to me, who am permitted to see only the impersonal material that is fed into the box and that issues from it? A particularly important operation, which is a kind of hybrid between instrumental and paper and pencil operations, is the "mental experiment."

Analysis of all that can be discovered in "mental" operations is obviously of a profitless and impossible complexity, but continued practice of the analysis makes me continually more conscious of the importance of the "verbal" aspect of what I do. There is no sharp distinction between "paper and pencil" and "mathematics" and "verbalizing" in general. Mathematics is merely a precise verbalization; we recall the aphorism of Willard Gibbs, "But mathematics is also a language." At first flush, insistence on the importance of the verbal must appear as entirely banal. Surely the analytically minded public has been made conscious of the importance of the verbal factor in its habits of thought through a host of recent writings, to mention only Korzybski's *Science and Sanity* and the recent *Tyranny of Words* of Stuart Chase. These writings, however, have been mostly directed to showing the dangers concealed in an uncritical acceptance of traditional habits of verbalization. This campaign has been so successful that "verbal" and "verbalization" have become epithets of reproach. It is the other aspect of the picture that I am concerned to emphasize here, namely that our traditional verbal habits may have the highest guiding and constructive value.

The concepts that we actually use in physics and the operations which give them meaning are only a few of the enormous number of conceivable concepts and operations. It would be most interesting to make a study of the properties of the concepts and operations which have survived, to find what sets them off as more suitable and convenient than any others conceivable. This study has not, so far as I am aware, ever been attempted. If it were made, I am convinced that the verbal factor would be found to be one of the most important factors determining selection and survival. The concepts of physics which we inherited ready-made were such that they fit the same verbal pattern as the more familiar objects of daily life; we demand of any new concepts which we are forced to invent to meet new previously unknown situations that they permit the familiar verbal handling.

It is no accident that so many times we are able, by giving heed merely to our verbal demands, to evolve a concept or point of view that is relevant to an "external" physical situation. For our verbal habits have evolved from millions of years of searching for adequate methods of dealing verbally with external situations, eliminating methods that were not a close enough fit. The desirability of continuing to use our old verbal habits in new situations if possible is obvious enough in achieving economy of mental effort, and the probability of at least a partial success is suggested by our universal experience that absolutely sharp breaks never occur, but that a method, hitherto adequate, can always be extrapolated beyond its present range with some partial validity. By the same token, however, the validity of any extrapolation may be expected eventually to break down, so that one may anticipate ineptnesses or inadequacies in concepts which have been formed by too uncritical a verbal extrapolation.

It would be interesting and perhaps profitable to reexamine all the concepts of physics which I analyzed in the *Logic of Modern Physics* from the verbal point of view, to find if possible just what part the verbal played historically in the evolution of the concept, and to what extent the utility of the concept has been circumscribed by its origin, or even to what extent the concept is commonly misused because of its verbal bar sinister. Such a program, however, would probably be tedious, and it is in any event more ambitious than I am in a position to attempt. I believe the purpose can be well enough accomplished by something less ambitious, which at the same time gives me a chance to attempt another thing in which I have long been interested. Perhaps the least adequate analysis which I attempted in the *Logic of Modern Physics* or in any of my subsequent writings was of the concepts of thermodynamics. There is a curious inconsistency between the attitude of most physicists toward thermodynamics and their attitude toward kinetic theory and statistical mechanics into which I have long wanted to probe. The interest of such a probe has increased rather than decreased in the last years since the concepts of probability have come to play such a fundamental role in wave mechanics. In the following pages I shall therefore try to kill two birds with one stone: namely to make an operational analysis of the situation in thermodynamics and related fields, and at the same time cast the analysis in such a form as to bring out the role and the consequences of the verbal requirements which have entered into forming the concepts. I shall attempt especially to make clear whether the operations involved are instrumental operations or "paper and pencil" operations; I have been often troubled in trying to understand expositions, particularly of modern physics, by this sort of ambiguity as to the nature of the fundamental operations.

THE NATURE OF THERMODYNAMICS

CHAPTER I

THE FIRST LAW OF THERMODYNAMICS AND THE CONCEPT OF ENERGY

IN SOME of his more exalted moments the physicist ascribes to the first and second laws of thermodynamics the most far-reaching generality he has ever achieved in formulating laws of nature, while on other grayer mornings he depreciates the insight attainable by thermodynamic analysis, extols the superior virtues of statistical mechanics and kinetic theory, and may even go so far as to maintain that the existence of phenomena like the Brownian fluctuations is positively inconsistent with the truth of thermodynamics. The latter mood seems to me by far the more common. There can be no doubt, I think, that the average physicist is made a little uncomfortable by thermodynamics. He is suspicious of its ostensible generality, and he doesn't quite see how anybody has a right to expect to achieve that kind of generality. He finds much more congenial the approach of statistical mechanics, with its analysis reaching into the details of those microscopic processes which in their larger aggregates constitute the subject matter of thermodynamics. He feels, rightly or wrongly, that by the methods of statistical mechanics and kinetic theory he has achieved a deeper insight.

It must be admitted, I think, that the laws of thermodynamics have a different feel from most of the other laws of the physicist. There is something more palpably verbal about them – they smell more of their human origin.

The guiding motif is strange to most of physics: namely, a capitalizing of the universal failure of human beings to construct perpetual motion machines of either the first or the second kind. Why should we expect nature to be interested either positively or negatively in the purposes of human beings, particularly purposes of such an unblushingly economic tinge? Or why should we expect that a formulation of regularities which we observe when we try to achieve these purposes should have a significance wider than the reach of the purposes themselves? The whole thing strikes one rather as a verbal *tour de force*, as an attempt to take the citadel by surprise.

We usually do not proceed like this in other fields. The most important comparable situation is that with regard to the restricted principle of relativity formulated by Einstein. Here we have also wrung a universal principle out of a universal failure of ours, this time our failure to make an experiment the results of which we would be willing to describe as measuring our velocity with respect to the ether. Many physicists, I suspect, feel about the principle of relativity much as they feel about thermodynamics: they do not see why the thing should work; there is a lack of perspicuousness about it that leaves them uncomfortable.

There are other features involved in the uncongeniality which so many physicists feel in the presence of thermodynamics and which makes it the most difficult branch of physics to teach. An important factor, I think, is purely formal: thermodynamics uses an unfamiliar brand of mathematics, and this often makes it difficult to find the method of formulating a problem that will lead to the desired result. It is a common complaint that there seems no way of hitting on a "cycle" that one can be sure in advance will disgorge the desired information. Difficulties of this kind I think are quite superficial and can be overcome by more care in teaching methods, but there are other deeper difficulties. For instance, why is thermodynamics restricted to the formulation of necessary conditions, and why is it so impotent in its endeavor to frame sufficient conditions? Other branches of physics are not thus restricted. Or why is it that it is so impotent to deal with irreversible processes? There are certain irreversible processes that are of a patent simplicity, and that can be completely measured by the instruments which give us our thermodynamic information, such as thermal conductivity. Why should not the physicist be able to deal with the thermodynamic implications of thermal conduction?

A part of the present difficulty and uneasiness I believe is connected with the change in the experimental situation since the time of the great founders of thermodynamics. Kelvin and Clausius were not acquainted with the fluctuation phenomena now familiar to everyone in the Brownian motion or in the small "shot" effect so readily demonstrated with a loud speaker and a vacuum tube. It can at least be argued that a knowledge of these phenomena would have vitally affected the turn which their deliberations took. As it was, Kelvin was obviously uncomfortable about the whole matter when he formulated the second law in terms of "inanimate, material, agency," a restriction so surprising as to be almost an admission of defeat. The difficulty was emphasized further by Maxwell and his demon, and was unsatisfactorily met by the pious hope that for some inscrutable reason no demon would ever be able to crash the gate of our laboratories. But today, when it is so easy to conjure the capricious happenings of the atomic world up into the control of events on the scale of daily life by Geiger tubes and amplifiers, I believe that many physicists honestly do not know whether or not to think that a sufficiently ingenious combination of means now in our control might violate the second law on a commercially profitable scale.

Thermodynamics gives me two strong impressions: first of a subject not yet complete or at least of one whose ultimate possibilities have not yet been explored, so that perhaps there may still be further generalizations awaiting discovery; and secondly and even more strongly as a subject whose fundamental and elementary operations have never been subject to an adequate analysis.

The average physicist perhaps feels these difficulties to a certain extent, but he feels also, I believe, that they could be adequately met by the methods of statistical mechanics and kinetic theory. There are considerations, however, which I believe should be a little disturbing to the complacency with which the possibility of a final statistical-kinetic solution is accepted. The structure of kinetic theory is built on the familiar mechanics of the objects of our everyday experience. We assume that the laws and concepts of Newtonian mechanics, proved by experiment to be valid for the objects of daily life, continue to have their same meaning and to be valid when extended to objects so small that the direct experimental check has never been given, and it is questionable whether even the individuality of the particles has the same meaning as on a larger scale. Even if the Heisenberg principle allows the ordinary mechanical parameters of any particular particle to be determined with the precision demanded by the kinetic analysis, it is not at all evident that the parameters of *all* the particles could be simultaneously determined with the required precision. Even if such a determination is permissible "in principle," no one in his wildest moments would tackle the job of making an instrumental determination of the parameters, but demands only that he be allowed to write on paper symbols representing the parameters of all the particles. That is to say, the meaning which can be ascribed to the parameters used in the analysis, which gives the physicist such a warm feeling of physical insight, has ceased entirely to be an instrumental meaning and has become purely a "paper and pencil" meaning. Why is the physicist so pleased with this? Why is he so willing to extrapolate laws beyond the reach of verifiability or of ordinary meaning, and why does he feel that he has achieved an "explanation" by so doing? Of course he still feels familiar and at home with the paper and pencil extensions of his ordinary experience, and he finds in the results of manipulations with these extensions the counterpart of the phenomena which he is trying to understand. But what shall he say if we ask him how he knows that there are not other methods of making the extensions and other methods of manipulation which would yield him the same counterpart? for if there are other methods, which one is it that gives the "right" explanation? Or if he replies that the "rightness" of an explanation is meaningless in the conditions, and we will have to grant the cogency of the objection, what confidence can he have that the particular extension which he has hit on will continue to be valid in new circumstances? We might also bring up the logical difficulties connected with the application of probability concepts to specific happenings, difficulties which I feel to be unsolvable.

Our inability to so manipulate our large-scale instruments as to get results that we think we could get if we were in possession of small-scale instruments I feel still eludes satisfactory understanding. The insight which the probability interpretation of the second law at first seemed to give turns to ashes like apples of Sodom. The small-scale stuff is only a model, obtained by extrapolation of the large-scale stuff. The tactics of this extrapolation certainly cannot be claimed to display any subtlety – it was what any child might have invented. The only check on the extrapolation is that when worked backward it shall again produce the large-scale stuff. One is tempted to think that there must be some other way of extrapolating to the small, in which all the mechanical phenomena of ordinary experience are not reproduced as perfect little conceptual replicas of themselves, but there is something different. Of course the trouble in searching for another sort of extrapolation is that the function of the extrapolation is to explain, and explanation always involves the reduction to something familiar. Any other sort of extrapolation would have to face the reproach that it was an *ad hoc* construction. Wave mechanics perhaps is such another way of extrapolating; a probability calculated as a number of ways of distributing energy among the solutions of Schroedinger's equation has a different ring from a probability calculated from a distribution among hard point atoms. But it is known that the meaning of all the concepts of wave mechanics is ultimately to be found in macroscopic operations – the classical difficulty crops up again, although in a slightly different guise.

The real reason the physicist is so satisfied with statistical mechanics and kinetic theory is that it works, and furthermore it works without making too great a tax on his manipulatory abilities. But knowing what we do now of the methods of physics, we have no right to anticipate that such a procedure must of necessity work, whatever may have been the optimistic expectations of our fathers in the less sophisticated days when statistical theory was first put together. We have no right to expect that it will work until we have tried it, and in finding that it does work we have made an empirical discovery of great importance. But the "understanding" that we set out to achieve proves to be an illusion; instead of understanding we have substituted a new discovery, which in turn awakens again the same old human craving for understanding.

It is certainly a paradox that the two laws to which the physicist ascribes the most sweeping universality, the laws of the conservation of energy and of the inexorable increase of entropy, are simply the first and second laws of *thermodynamics*. At least the first of these is used explicitly in kinetic theory, and the ideas of the second are now interwoven into the very foundations of quantum mechanics. By what magic has our stream risen higher than its source?

An understanding of the attitude of physicists toward thermodynamics and kinetic theory is, I think, to be sought only in the realm of psychology. Ever since the days of the Greek philosophers or of Lucretius human speculation has run straight to the atomic. At first there was absolutely no experimental justification for this, or logical justification either, for that matter. From our present point of vantage we must not draw the conclusion that because atoms have now been found in the laboratory our primitive urge to analyze into atoms was therefore justified. It just seems to be a fact about our thinking machinery that we must have our atoms; we cannot think of the velocity of a uniform fluid without imagining the "particles" of which the fluid is composed, nor can the most advanced speculations of quantum mechanics do without their ultimate particles. Perhaps the human necessity for its particles is connected with the necessity for "identification" in thinking; the very words we use are little islands of identification in the amorphous sea of our cerebration, and the particle of a fluid is a little materialized piece of identifiability.

TEMPERATURE

We may now start on a detailed examination of what thermodynamics involves. We shall not assume a fictitious ignorance and attempt an arrangement and development of the fundamental ideas in a logical hierarchy, but shall assume the background of present-day man. We begin with an examination of the concept of temperature. We may take temperature as roughly characteristic of thermodynamic systems, and distinguish a thermodynamic analysis from a non-thermodynamic analysis by the occurrence of the term "temperature" in an important role. Our first task is to find what it is that we do when we treat phenomena involving temperature in such a way that experience leads us to anticipate success. Approached in this way, it is obvious enough that the actual situation is of the greatest complexity, and that the steps in our successful manipulations have grown up by a process of spiraling approximation which we would find it very difficult to break down cleanly into independent factors in such a way that we could make it the basis of a postulational formulation. The starting point is the undoubted qualitative connection between the temperature concept and our crude physiological sensations of hot and cold. But there is also the undoubted fact that the physical manipulations with which we are most familiar and which we regard as most important ignore the temperature factor to a first approximation. The temperature of the environment of ordinary experience does not vary through a very wide range. When we specify the operation by which we measure length we pay no attention at first to the temperature at which the measurement is made, and furthermore to a first rough approximation we think of the length of ordinary objects as unaffected by temperature. In the way in which it has grown up and is used, temperature is, by and large, a rather secondary thing, and we do not bother with it until we begin to make refinements in our other methods of dealing with objects.

The main facts with respect to the attainment of temperature equilibrium were long known: a tumbler of hot water and one of ice water left in a room over night have in the morning come to the same "temperature," which is also the "temperature" of all the other objects in the room. We first analyze what we can do in a large room which has been standing long enough to come to "temperature" equality. This does not mean that we are necessarily going to rule out operations in the room which may give rise to temperature differences, such as rubbing a button on a board. Neither does it mean that we are going to ignore the fact that at different times of the year the temperature of our laboratory room will be different, and we are even willing to admit the possession of some sort of thermometer by which we can describe the variations of temperature from day to day.

Assume now our laboratory room all at constant temperature. Our ultimate problem is to acquire complete mastery of everything in the room; this means at least the ability to describe exhaustively everything that happens to us, and to anticipate everything that is going to happen. To achieve this mastery we make certain analytic breakdowns. First of all, a million years of experience leads us to recognize the presence in the room of various "bodies." We need not stop for an analysis of what we mean by saying that there are "bodies" in the environment, but shall assume that this is sufficiently obvious. For instance, the idea of "body" involves a certain identifiability and permanence. We first consider the bodies "in themselves," that is, as fixed and definite things, isolated from each other and not moving with respect to each other, and we try to characterize them adequately in their immobility. What we get in this way we usually describe as the "properties" of the body. A precise analysis of what is involved here is evidently tremendously complicated, and we shall not attempt it, assuming that the essential features of the situation are already well understood from previous applications of the operational method. A complete characterization of the body would demand the ability to specify what will be the result of every conceivable operation applied to the body. The "properties" of the body are more special and are loosely understood to be the results of operations under which the body is as a whole inert or passive, so that it is "unaffected" by the operations. Almost at once we discover that the "properties" of a body are not always the same but change when the external circumstances of the body change, as the density of a quantity of gas changes when the pressure is altered. The empirical result then soon appears that the different properties of any concrete body are not capable of *arbitrary* variation, but they are tied together, so that when certain properties have assigned values all others are uniquely fixed. Thus when the pressure of a certain quantity of gas is fixed under certain simple conditions at the temperature of our laboratory, its density, viscosity, thermal conductivity, dielectric constant, etc., etc., are fixed also. This means that density, viscosity, etc., can be expressed as functions of the pressure, which mathematically plays the role of an independent variable. But by a transposition, pressure may be treated as a function of density, so that density might be used as the independent variable, and pressure, viscosity, etc., become functions of it. It is the *number* of independent variables that is important. A more complete examination of the gas might well reveal the existence of more than one independent variable, such for example as the magnetic field. The *qualitative* aspect of our mathematical treatment is therefore a reflection of the precision of our measurements, since we might change from functions of one variable to functions of two variables merely by increasing the sensitiveness of an instrument. Such a situation is by no means novel.

Imagine that we have adequately determined all the properties of the bodies, in particular the relation between pressure and volume, in our laboratory on some day in winter. If we make the same observations (we assume that it does not bother us to assign a meaning to "same" under these conditions) on a day in summer, the relation fails, and volume, thermal conductivity, etc., become new functions of pressure. Considering all such measurements on all the days of the year, we may say that density is a function of pressure, the function itself being a function of a parameter that specifies the day of the year. We then notice that the new function has the same value on all days of the year that have the same temperature as indicated by our physiological sensations or by whatever crude thermometers we are sophisticated enough to construct. Hence eventually we see that under all conditions the properties of the body are determined in terms of one more than the original number of independent variables. By a mathematical transposition this extra variable may be chosen as any one of the original dependent variables – volume, for instance, for a simple gas. This gives eventually the possibility of constructing a thermometer by exploiting any one of the original "secondary" properties of the substance, such as volume, and thus ultimately of getting away entirely from any reference to our physiological sensations of hot or cold in framing the temperature concept. This is all so well understood that more detailed discussion is hardly necessary. The important thing is that only *one* variable in addition to our original set is necessary.

What is the process by which one assures himself that he has measured the true temperature of a body? Probably it is not essentially different from that used with any other sort of measurement. The factor of repetition is involved in the first place; we have to be able to repeat measurements on what we believe to be identical systems, and the only rigorous criterion to show that they are identical is that all their past history has been the same. Given a repeatable system, then, one makes measurements on it with a series of thermometers of different sizes, with different absorbing coatings on the bulb, and maintained in contact with the system for different times. One demands that in order to get the true temperature the readings must approach a smooth limit for thermometers below a certain size and for times of contact less than a certain upper limit but greater than a certain lower limit specifiable in advance for each thermometer and perhaps varying with the nature of the system. The question arises whether a reading will always be attained that can be called temperature. That is to say, in addition to bodies in which the temperature varies from point to point and with the time – which will be discussed later – can the temperature concept also be applied if there is a strong radiational field, or if the nature of the system is itself slowly changing with time, such for example as a change in the relative proportions of para- and ortho- hydrogen? With regard to this sort of internal change the position must be that no case has yet arisen in which it has not been possible to devise an instrument or a procedure such that an independent parameter can be assigned to the amount of internal change, the instrument operating only on the material as given and not demanding a knowledge of its past history. In the case of para- and ortho- hydrogen, for example, the instrument for determining the instantaneous composition of the system might be a Stern-Gerlach apparatus.

With regards to systems which are not in equilibrium because they may be in a uni-directional stream of radiation, or bathed in radiation of one wave length, the situation is not so clear. There may be cases in which a system shows steady behavior, but in which thermodynamic equilibrium is by no means attained, as when we speak of the electrons in certain vacuum-tube phenomena being at temperatures of thousands of degrees. The question arises as to how one characterizes a radiational field. This question is dealt with in Planck's well-known book on radiation. I do not believe, however, that the treatment there is entirely satisfactory; there is no specification of the measuring instruments, and it is assumed that the same sort of instrumental manipulation that can be performed on the macroscopic scale can be carried down into the microscopic. The spirit of Planck's treatment is statistical rather than thermodynamic. However, from the numerical values given it appears why one does not ordinarily bother with radiation in thermodynamic arguments. (Strictly speaking things would be invisible in a black-body radiation field, and no observations could be made.) At room temperature the density of black-body radiation is 1.4×10^{-5} ergs/cm^3. Since the specific heat of water is $4.18\ 10^7$ ergs/gm°C, the total radiational content of one gram of water at room temperature would raise its temperature by something of the order of 10^{-12} °C. Further, the ordinary radiation fields in which bodies are bathed and by which we are able to observe them involve vanishingly small amounts of energy. The radiational intensity of full sunlight is 1.93 cal/min cm^2, which means a radiational intensity of $\frac{1.93}{60 \times 3 \times 10^{10}} \approx 10^{-12}$ cal/cm^3, or again about that required to raise one gram of water 10^{-12} °C.

In spite of the smallness of the quantities involved, we ought to have some paper and pencil criterion for determining whether there is an equilibrium radiation field in the interior of a body. Perhaps this sort of thing would do: Imagine perfectly reflecting or perfectly absorbing surfaces of zero mass and zero heat capacity, freely movable in the body. Then if the radiational field is the black-body field characteristic of the temperature of the body, the reading of a thermometer inside the body must be unchanged irrespective of how these surfaces are moved about.

A body whose properties all have specified values is said to be in a certain "state," and the independent variables which are chosen to specify the properties are called the "parameters of state." Functional connections between the state parameters and the other dependent variables are given by various "equations of state." It is sometimes intimated that the notion of "state parameter" can only be expressed in terms of the laws of thermodynamics, as for instance, the parameters of state are those in terms of which the internal energy is a complete differential. This, it seems to me, incorrectly reproduces what we actually do. The notion of "state" is prior to the laws of thermodynamics; in fact if it were not prior the first law of thermodynamics would be reduced to a tautology. It is true, however, that we do have to ask with some care to what extent the notion of "state" actually applies to the bodies of our experience. Do bodies have "properties" such that when they are fixed the behavior of the bodies is fixed? The answer is often simply enough "Yes," and it is to such bodies that our thermodynamics applies, but there are situations where this treatment is not so obviously adequate, as whenever we have a body exhibiting hysteresis.

Having now our bodies with their properties we allow them to interact with each other or to undergo any changes of which they are capable, and our problem is to acquire mastery of this extended range of phenomena. We again effect our first simplification by studying the changes which take place in our laboratory room all at the same temperature. So far as the phenomena do not involve the spontaneous appearance of temperature differences, as for example generation of heat by friction or collision, we assume for the purposes of our thermodynamic analysis that we are already completely master of the phenomena. This will doubtless involve the introduction of new variables, such as the velocity of motion of parts of the system. All the variables which are necessary under these conditions we may lump together under the term "mechanical" variables; they include what would ordinarily be called electric or magnetic variables and similar others. The implication of "mechanical" is simply "non-thermal." When we assume that we have complete mastery of phenomena at uniform temperature, this means in particular that we know all about the "mechanical" forces in their generalized sense, and are able to express the "work" done by these forces in any generalized displacements. Our assumed mastery of all non-thermal phenomena corresponds to various subdivisions of physics which are treated separately in textbooks, such as electrodynamics or optics.

Having mastery of the "mechanical" aspects of phenomena, the task of thermodynamics is to give us mastery of what happens when the "thermal" factor plays a role, that is, when temperature differences appear in the system or motions appear in virtue of temperature differences. A thesis of thermodynamics is that certain very broad aspects of the new and extended phenomena can be exhaustively handled by the introduction of the temperature of the various parts of the system – the same variable already demanded and adequate for an exhaustive characterization of the "properties."

Such is the broad background. One cannot get very far, however, in analyzing what is involved without seeing that for the purposes of thermodynamics greater precision is necessary than is ordinarily considered to suffice for the purpose of mechanics. It would be quixotic, however, to imagine that we can foresee all necessary refinements in a vacuum, and the best course will be to leave the refinements until we actually need them when we come to discuss the limitations of the first and second laws of thermodynamics.

In the meantime we shall make all the naïve assumptions that are usually made about thermal phenomena; we shall assume that bodies may *have* definite temperatures, that these may be read by thermometers if we wait long enough, that two bodies each at the same temperature as a third are at the same temperature as each other, etc., etc. All these come from our fundamental assumption that the contents of a large room will come to equilibrium at a single temperature if we wait long enough. In addition to these fundamental assumptions about temperature equilibrium we assume the usual familiarity with the facts of calorimetry.

CERTAIN ASPECTS OF CALORIMETRY

A logical analysis of the concepts which are based on these facts might be well worth while, but is, I believe, hardly necessary for our purpose. We may profitably stop, however, for a few comments on some of the aspects of the concepts of calorimetry.

It is well known that in the historical development the treatment of heat as a "thing" played a most important role. It was later found that heat is not in all respects like a "thing," but we may nevertheless make an experimental study of the conditions under which it is "transferred" from one body to another. We find two qualitatively different sorts of condition under which such a transfer takes place. In the first place, the bodies may be in actual contact. It is an experimental fact that under such conditions there is no temperature discontinuity at the actual surface of contact, assuming of course that there is sufficient homogeneity in condition so that the temperature concept is applicable. When the bodies are thus in contact it is an experimental fact that there is a transfer of heat if there is a temperature gradient in the material of the two bodies in the directions normal to the surface of separation. This sort of heat transfer is said to be by conduction. There is a property of the body, in general a function of its temperature (and perhaps mechanical variables in the general case), which may be determined numerically as a coefficient, such that the heat transferred by conduction per unit area per unit time is the product of this coefficient and the temperature gradient.

The second method by which heat transfer may take place occurs when two bodies are not in contact, but confront each other across a vacuous space, with their opposing surfaces at a difference of temperature; this method of transfer is "radiational" transfer. Again the amount of heat so transferred may be determined by measurements and calculations made now at the places involved. The situation is, however, a little more complicated than it is with regard to conduction. New characteristic coefficients appear associated with the *surfaces* of each body, namely coefficients of emission and of absorption and of reflection, such that the total heat transferred from the one body to the other is given by a not-too-complicated mathematical expression in the temperatures of the opposing surfaces and the coefficients of each surface. These coefficients are determinable so as to reproduce the measured heat transfer between all possible pairs of substance under all conditions of temperature difference. Although the situation is more complicated than before, the multiplicity of combinations is an infinity of a much higher order than that of the coefficients to be determined, so that the coefficients are amply determined and at the same time the underlying assumptions can be checked and given meaning. The coefficients thus determine *properties* of the surfaces.

Not only may bodies exchange heat by radiation across a vacuous intervening space, but they may also radiate to each other through a separating layer of matter, provided that matter is "transparent." Deficiency in transparency means an absorption in the interior of the matter and a corresponding reëmission. We therefore arrive at three new coefficients to characterize the interior of a body in addition to the thermal conductivity which we have already. The three coefficients, transparency, absorption, and emissivity, are to be associated with the radiational process as distinguished from the conduction process.

We now have to ask whether this picture to which we have been led by our calculations corresponds to anything real. When we determine a coefficient of conductivity by the usual measurements in terms of heat transfer and temperature gradient are we not including at the same time the heat transferred through the body of the substance by "radiation," and is there any physical operation which we can perform to separate the two processes of transfer? This question is seldom if ever raised in discussions of thermal conductivity. If we are dealing with large amounts of substance, I think there is no operational method of separating the two mechanisms, for if we double the temperature gradient everywhere, we have simultaneously doubled the temperature difference between any two points, so that we have simultaneously doubled the heat transferred by "conduction" and by "radiation." The distinction is operationally possible only when we pass to layers which are thin and whose thinness is of the right order of magnitude. Imagine a layer of the substance separating two surfaces maintained at a definite temperature difference. The thickness of the layer is to be great enough so that the temperature may be determined at every point of its interior with microscopic thermometers, so that the temperature gradient everywhere has a physical meaning, but the thickness at the same time is to be small enough so that some of the heat which leaves one of the surfaces by radiation passes through the layer without complete absorption and impinges directly on the other surface. Under these conditions, by sufficiently varying the conditions, such as the thickness of the layer and the emissivity and absorption of the two surfaces, it will be possible by calculation to separate the transfer into its two parts, since the two parts depend in different ways on the variables, and thus to give meaning to the two processes. If, however, the opacity of the substance is so high that volume absorption is complete in a distance so small that physical meaning cannot be given to the temperature of the corresponding volume element because of its smallness, then the two processes are not operationally separable, and it is meaningless to talk of two.

The reason that this matter is not discussed in elementary texts in connection with the measurement of thermal conductivity is doubtless that under ordinary conditions the radiational contribution is presumably negligibly small in comparison with the conductive. I have not, however, seen the actual calculations of this point. But at high temperatures, as in the interior of the stars, the radiational transfer becomes important, and the experimental determination of the coefficient of conductivity under such conditions would become much more complicated than it ordinarily is, although the corresponding pencil and paper operations remain simple enough.

Going back to the concept of heat as a "thing," the discovery that finally made physicists discard this point of view was that "heat" is not conserved, but disappears when mechanical work is done. Another consideration might have shown the unsuitability of this view before the enunciation of the first law, namely that it is not possible to analyze the flow of heat into the product of a density and a velocity, something which we can always do when we are dealing with the flow of matter. The whole situation with regard to flow is evidently embarrassed by strong verbal impulses: we find it difficult to say "There is flux at this point" without wanting to say also "Something is flowing at this point," and if something, then there is a velocity and a density. To what extent is the impulse verbal that demands a velocity if there is a "thing"?

FORMULATION OF THE FIRST LAW

The first law of thermodynamics is usually written in the conventional differential form

$$dE = dW + dQ.$$

Here dQ is said to be the heat received, dW is the work received, and dE is the increase of internal energy, all in a definite interval of time. We ask what are the more precise meanings? In first particular, to what do the equations apply? The law is used with the widest possible generality; although it is perhaps usual to think of the application as made to some ordinary "body," it is not actually restricted to any definite piece of what might be described as a material substance, but is also applicable to such a thing as any part of an electromagnetic field, or to regions in which the content of material substance is continually changing, as when it is applied in hydrodynamics. That is, the equation is applicable to a "region," whether occupied by "matter" or not, and this region is defined by a surface which separates it from its surroundings. This surface may itself be in process of distortion. The equation has to do with what happens to the region as defined by its bounding surface in a certain interval of time. This time does not enter the formulation of the law explicitly, so that the implication is that the formulation holds no matter what the interval of time, whether large or small. This, I think, will be assented to. In order, however, not to get into complications which do not concern us, we assume that the region is small enough so that it all has the same "local" time, and that we do not have to consider propagation effects from one part of the region to another (that is, we shall not concern ourselves with relativity effects). There is, I think, no reason to anticipate that any special considerations necessary to treat propagation effects will react back to modify the concepts by which local phenomena are treated.

It is perhaps not customary in formulations of the first law to insist so articulately that the application is to a region separated from its surroundings by a boundary surface. When carried through to its logical verbal or paper and pencil conclusion this has the important result that we will be driven to ascribe full and complete localization to the "energy," for it is not possible to set up any consistent restriction on the way in which the surface may be drawn, but the law must apply no matter what the location of the surface. This is contrary to views frequently expressed with regard to localization of energy, for it is often said that energy is a property only of a system as a whole, and one must not inquire what may be the details of its localization. In fact, on page 111 of my *Logic of Modern Physics* I expressed much this view with regard to localization. However I can see no escape from the position which I am now adopting. It not only seems to be demanded by verbal consistency, but also is in accord with experiment, because it proves possible in every physical situation to localize an "energy" inside every surface no matter how drawn, and furthermore to give the dE, dQ, and dW associated with such an arbitrary surface full instrumental significance. This is possible even in the gravitational case, and without introducing any of the notions of Einstein's generalized relativity theory, contrary to a widespread opinion, as will be shown later. The following exposition revolves to a certain extent around this thesis that the first law may be formulated for any arbitrary closed surface in terms of the results of instrumental operations of one sort performed in the interior and of other sorts at all points of the surface. Although it may not be usual to make so explicit a statement, I believe that this is at least implicit in what physicists do when they handle the first law in its full generality, and it also is in accord with experiment.

Since, then, dQ, dW, and dE are associated with what happens to any region in any arbitrary interval of time, there is no reason except traditional reasons of manipulatory convenience why the equation should be written in differential form. Since our application is by no means to be restricted to infinitesimal intervals of time or to infinitesimal elements of volume, it seems best to write the law at once in finite form:

$$\left(\begin{array}{l}\text{"Gain of internal energy" of}\\ \text{region in any given time in-}\\ \text{terval}\end{array}\right) = \left(\begin{array}{l}\text{Work received by the region}\\ \text{from its surroundings in the}\\ \text{given time interval}\end{array}\right)$$

$$\text{plus}$$

$$\left(\begin{array}{l}\text{Heat received by the region}\\ \text{from its surroundings in the}\\ \text{given time interval}\end{array}\right)$$

This may be written symbolically:

$$\Delta E = W + Q$$

What now are the operations by which we get the Q in any concrete case? Q is the net heat which enters the region. This heat gets into the region across the bounding surface from the outside. To obtain its numerical value we have to imagine sentries posted at every point of the boundary equipped each with his instrument by which he measures the amount of heat that flows past his element of surface, and then the contributions recorded by each sentry are to be added at some central clearing-house in order to get the total net Q. For a region of infinitesimal size bounded by a stationary surface and for unit time interval the clearing-house operations obviously give the product of volume and the negative divergence of the heat flow. For a region of unit size and for unit time interval we may write $Q = -\operatorname{Div} \vec{q}$, where $\vec{q}$ is the vector heat flow.

The situation is thus peculiar; the instrumental operations are in terms of a flow vector which can be measured at any point, whereas the pencil and paper operations, which give the first law its meaning and to which the instrumental operations are preliminary, demand less than a full instrumental knowledge, being satisfied with merely the divergence of the flow vector. There is, I think, no simple instrument which responds directly to the *divergence* of heat flow. The heat flow at every point, as far as the first law is concerned, might be modified by the addition of any arbitrary flow, no matter how large, or how completely without instrumental significance, provided only that it is divergenceless.

By writing the first law in terms of Q or Div $\vec{q}$, we have obviously assumed that we know how to measure heat flow, or $\vec{q}$, at a point. This of course is carried over from our assumption that we are masters of calorimetry. How are the instruments constructed by which we measure heat flow in calorimetry? I believe that there is no simple instrument for doing this, any more than there is for measuring Div $\vec{q}$, but in any actual case we get the $\vec{q}$ by a more or less complicated process, involving both instrumental and paper and pencil operations. Apparently the simplest way is to determine the temperature gradient at the point in question by infinitesimal thermometers an infinitesimal distance apart, and then to calculate the heat flow by combining the gradient with the thermal conductivity of the material, which we have had to determine by some suitable independent preliminary experiment. We are not proceeding in a circle here, as at first might appear, because thermal conductivity and quantity of heat may eventually be defined in terms of temperature drops of definite pieces of matter and not in terms of a vector flow. A quantity of heat is always a quantity which has left or entered some body.

Conventional calorimetry recognizes at least two ways in which the heat which has left a body during a process without work may be measured. The common way involves making the body lose heat to an infinite reservoir. By making it infinite, any rise of temperature of the reservoir during a finite process is made infinitesimal, but at the same time the only way of measuring the heat which has left the body is to measure the infinitesimal rise of temperature of the reservoir and multiply by its infinite heat capacity. This is physically inelegant and mathematically anathema, involving as it does the necessity for physical instruments capable of dealing with "infinitesimals" of different orders; perhaps however it is not logically unrigorous. It seems to me that a second method is somewhat better, namely to make the heat reservoirs take the form of ice calorimeters. A finite loss of heat is then accompanied by a finite loss of volume, or change of phase, and there is no need of any complication by the entrance of infinity. If one wants to be academic enough to object to the finite number of substances that can be used in calorimeters in this way, and hence the finite number of reference temperatures available, one can make the temperature range continuous by subjecting the calorimeters to variable pressures, taking pains to apply the pressure to the piston at some other surface than that at which heat transfer takes place in order to avoid complicating the heat transfer with mechanical work. Even the ice calorimeter must have a certain temperature gradient at the surface in order that heat may be transferred at all, but disturbances of this sort can be made to vanish to any necessary degree.

Another question which might cause uneasiness in determining heat flow is with regard to measuring the temperature gradient at a point in terms of the temperature difference of two infinitesimal thermometers an infinitesimal distance apart, for surely this is not a measurement "at a point." I do not believe, however, that this question is particularly important, any more than the corresponding question with regard to measuring the velocity of any ordinary body in terms of two successive measurements of position and calling the result the "velocity at an instant." It may be that some radically different method of measuring heat flow could be devised, such as determining the differential pressure on the two sides of a vane sensitive to the material elastic waves which constitute heat, in the same way that a tachometer can be made by measuring the centrifugal force of a governor arrangement.

In any event, whatever the details of the measurements, we do have to assume, if we wish to formulate the law as written, that the conditions demanded to make possible the ordinary operations of calorimetry still hold under the general conditions assumed by the first law. But this is almost certainly not always the case. An examination of what is involved in classical calorimetry shows that its operations are performed in the absence of ordinary mechanical phenomena, whereas here we have to know the meaning of heat flow across a surface which may be in motion and across which mechanical work is simultaneously being performed. So general a situation is certainly not contemplated in the original operations of calorimetry, and whether the original operations can be sufficiently broadened is to my mind very questionable. At least I have never seen any adequate discussion of the possibility. In the application of the first law to any concrete system we would do well, therefore, to make our set-up, if we can, such that the system receives no work across those portions of its bounding surface where it is receiving heat, and vice versa.

WORK AND THE FLOW OF MECHANICAL ENERGY

Turn now to an examination of the W of the first law. This W means the total mechanical work received by the region inside the boundary from the region outside. As in the case of Q, this work is done across the boundary, and the evaluation of W demands the posting of sentries at all points of the boundary, and the summing of their contributions. In the simple cases usually considered in elementary discussions the work received by the inside from the outside is of the simple sort typified by the motion of stretched cords or of simple linear piston rods. Our sentry can adequately report this sort of thing in terms of finite forces acting at points and finite displacements. In general, however, there will be contact of the material inside with the material outside over finite regions of the boundary, and we become involved in the stresses and strains of elasticity theory. Under these more general conditions it is not so convenient to speak of the action of one body on another in terms of work (a detailed discussion will be given presently of the infelicities that result when we apply the notion of work to the sliding of two bodies on each other with friction), but the situation is adequately covered by the introduction of a generalized Poynting vector. This generalized Poynting vector may be called simply the "flow vector for 'mechanical' energy." "Mechanical" is used here in the same sense that we have already employed in connection with "mechanical" parameters of state, namely as meaning merely "non-thermal." The introduction of this nomenclature at this stage is an anachronism from the point of view of an elegant exposition, because it assumes as already known the generalized energy concept to which we are entitled only after the complete establishment of the first law. But I am not particularly interested in maintaining the fiction of an innocence which I cannot claim, and the advantages of introducing the term "mechanical energy" at this stage are too great to resist.

The W of our formulation of the first law is then merely the net flow into the region of "mechanical energy," or, if we write the law for a region of unit size and for unit interval of time, W becomes $-\mathrm{Div}\, \vec{w}$, where $\vec{w}$ is the vector flow of mechanical energy, or the generalized Poynting vector. We have the same situation in general for $\vec{w}$ therefore that we have already analyzed for the flow of heat. Now $\vec{w}$ has unique instrumental significance because the sentries posted at the boundary can make instrumental determinations of $\vec{w}$ at every point (the detailed rules in some special cases will be considered presently), but it is only the divergence of $\vec{w}$, which has no immediate instrumental significance, which occurs in the first law.

In writing the first law in terms of $\vec{w}$ we make the same implication that we did with regard to the heat flow, namely that we are complete masters of the mechanics of the situation, and know how to determine the flow of mechanical energy at all points of the boundary under the general conditions contemplated by the first law. In general this assumes that we are able to give an adequate mechanical account of what is going on when thermal phenomena are simultaneously taking place. But we probably cannot do this in general for mechanical effects any better than we could for thermal effects; the practical answer is as before to isolate the two classes of phenomena when possible.

It will pay us to stop to look at this general notion of flow of mechanical energy because there are difficulties which are not usually appreciated. Probably the most frequent use made by physicists of the concept of flow of energy is in connection with the electromagnetic field, and in particular in connection with a pure radiation field. Here the relations are simple and lend themselves to a rather complete verbalization. The energy flow is given by the Poynting vector $S = c\frac{E \times H}{4\pi}$. This energy flow satisfies one very important criterion that we impose almost unconsciously as to its "physical reality," for the flow at any point is determined in terms only of instrument readings made at that point and at that instant, namely measurements of the electric and magnetic fields and of the angle between them. In other words, the satisfaction of this criterion gives the operational meaning of "physical reality" in this situation. We are encouraged by this to go further, and are accustomed to think of the flow of electromagnetic energy in a radiation field with all the rest of the imagery (that is, with all the verbalizations) which we apply to physically real *objects*. Consistent with this imagery we find that the energy flow can be analyzed into two factors: there is an energy density $\frac{E^2 + H^2}{8\pi}$, and the velocity with which energy of this density flows is $c\frac{2E \times H}{E^2 + H^2}$ namely the flow vector divided by the velocity. In the case of a pure radiation field in empty space this is the velocity of light, as it should be. Extension of the idea of the "physical reality" of the flow of electromagnetic energy to non-radiational fields does not always lead to results so happy. For instance, in the completely static field surrounding a single electrostatic charge and a single magnetic pole at some distance from it the Poynting vector indicates a perpetual circulation of electromagnetic energy in closed circles. Since, however, the problem of the transfer of energy seldom obtrudes itself in this sort of situation, we usually are not aware of the infelicity of pushing on to a verbal conclusion. Although this situation violates our verbal feeling of what is proper, it involves no consequences with regard to instrumental operations that are not in accord with experiment.

When it comes, however, to the flow of ordinary mechanical energy in an extended body which is the seat of ordinary elastic stresses, the situation is much less satisfactory. I think this is seldom realized, probably because most of the situations of ordinary mechanics are treated with sufficient completeness for most purposes by less elaborate methods. Consideration of an elementary example will show the points at issue. Imagine a simple mechanical motor, actuated by the unwinding of a spring, turning by means of a conventional pulley and belt the shaft of another machine which is nothing but a simple flywheel, accelerating under the drive of the motor. Draw two closed surfaces around the two machines. Then the mechanical energy localized inside the one surface is decreasing, as shown by measurements made with instruments inside that surface (that is to say, the spring is unwinding), and simultaneously the mechanical energy inside the other surface is increasing (the angular velocity of the flywheel is increasing). We verbalize this and say that mechanical energy is leaving one region by flowing out across its bounding surface, flowing across the intermediate space, and entering the other region through its bounding surface. If this flux of mechanical energy across the intermediate space is to be said to have "physical reality" we demand that it be able to manifest itself to physical instruments stationed in the intermediate space. The intermediate region may be divided into four different characteristic regions according to the sort of instrumental readings that can be obtained. In the first place, there are points of "empty" space; second, there are points in the taut half of the belt, traveling toward the motor; third, there are points in the slack half of the belt traveling away from the motor; and fourth, there are points in the common base to which both machines are attached, stationary, and with a total integrated compressive stress equal to the tensile stress integrated across the taut half of the belt. How shall we distribute the flux of mechanical energy among these four regions? In the first place, there can be no flux through empty space, because we know of no instrument which would give any readings at all in points of empty space in a purely mechanical system like this. This decision is easy; it will require a little more consideration to decide what to do with the other regions, because instruments situated in any of them do give positive readings. Consider the base to which the machines are bolted. Instruments disclose matter, at rest, with a compressive stress.

If we are to ascribe a flux of mechanical energy to this region, then we must postulate a flux of mechanical energy wherever else we find matter at rest supporting a compressive stress. But we can obviously set up completely static systems, in which matter is subject to a permanent state of compressive stress, in which we must postulate no mechanical energy flow by the principle of symmetry or of sufficient reason, if by no other argument. Hence we cannot locate our flux in the common base to which the two machines are bolted. Consider next the slack half of the belt. Instruments here disclose matter with no stress moving with constant velocity. If we are to locate mechanical energy flow here, then we must always have such flow when we have matter without stress moving with uniform velocity. But imagine an isolated section of the belt, with two free ends, moving uniformly through empty space. There is no stress in the material. The motion continues forever with no alteration in any of the measurable properties; we would obviously be going out of our way to postulate a flow of mechanical energy in such a situation. Hence we cannot locate the flux of mechanical energy in the slack half of the belt. There remains only the taut half, and unless we have been barking up the wrong tree, we must find our flux here. There is indeed no difficulty here, and the solution is obvious on inspection. We can get the proper transfer of mechanical energy by localizing in the belt a flux equal to the product of the tension and the velocity of motion of the belt. That is, the flux is obtained by a pencil and paper operation on the data of simple instruments, one of which gives the velocity of motion, and the other the tension. No particular significance is to be attached to the fact that we have found it convenient to mention a paper and pencil operation, for I suppose that any ingenious boy would undertake to construct an instrument which would automatically combine properly the readings of the two instruments.

The theory of elasticity permits our result to be generalized and a precise formula written for the flux of mechanical energy in any body in which there is a mechanical stress and which is moving in any way.

$$
\text{(A)}\qquad
\begin{aligned}
S_x &= -\,(\dot{u}X_x + \dot{v}X_y + \dot{w}Z_x)\\
S_y &= -\,(\dot{u}X_y + \dot{v}Y_y + \dot{w}Z_y)\\
S_z &= -\,(\dot{u}Z_x + \dot{v}Z_y + \dot{w}Z_z)
\end{aligned}
$$

where $\dot{u}$, $\dot{v}$, $\dot{w}$ are the velocity components, and the Y_z's the stresses.

We therefore retire from our first attempt to localize the flux of mechanical energy with considerable satisfaction, but it gives us a rather curious sensation as we glance over our shoulder for a last look at the problem to notice that the energy in the belt is running the "wrong" way, against the motion of the belt from motor to flywheel. But on reflection we do not see how it could possibly run the other way.

The obvious motivation of our verbal drives in handling this problem has been to be able to think of "mechanical energy" as a *thing*, and we have been successful up to a point, but our last discovery that the energy is flowing the "wrong" way may come with a considerable shock, and may start us to wondering whether after all we should be so pleased with our treatment of mechanical energy as a thing.

Another respect in which the correspondence with a thing is perhaps not as close as we were prepared to expect occurs to us almost immediately. There is no way of identifying pieces of mechanical energy; there is no operation by which I can give meaning to the statement "The energy which is now in this reservoir is the same energy as that which was at a previous moment in that reservoir." We usually regard the property of identifiability as perhaps the most essential characteristic of ordinary matter. But when we ask what we do in order to identify ordinary matter, and in particular homogeneous matter such as a fluid to which the equations of hydrodynamics apply, the situation is not so clear cut. The "particle" picture of matter is an invention which it has already been suggested was motivated by the intellectual necessity of being able to identify the parts of a homogeneous medium. For it appears that we usually do make this assumption of identifiability; the equations of hydrodynamics have meaning for most people, I think, only in terms of this assumption. The velocity that enters the equations is the velocity of an identifiable piece of the fluid. If we imagine the fluid composed of "particles," then the process of following an identifiable piece of the fluid is in thought merely the operation of following some selected particle. This of course is possible because by construction the particle has the property of identifiability. Later, in mechanics, we give the particle at least one other property, but we do not need to enter into a consideration of mass here.

The property of identifiability is thus perhaps the one which tradition has prepared us to think is the one essential of ordinary matter, and therefore it is at first disconcerting to discover that mechanical energy, which for convenience in thinking we would like to assimilate to ordinary matter, does not have this property of identifiability. On reflection, however, it seems possible that this may not be so vital a matter after all. Wave mechanics denies the property of identifiability or individuality to electrons or whatever other ultimate constructional units ordinary matter has, and it is the present conviction of many physicists that wave mechanics contains the potentialities of giving an adequate account of ordinary matter. It must be, therefore, that there is some alternative method of describing those properties of ordinary matter which scientific tradition has claimed involve identifiability, and we are accordingly not so disturbed that the operation of identification cannot be applied to energy. In fact it is not difficult to imagine an equivalent operation in the case of the perfect fluid of hydrodynamics. The velocity of such a fluid could be satisfactorily measured by measuring in the ordinary way the velocity of a microscopic vane immersed in the fluid, subject to the condition that there be no force acting on it. The measurement of the velocity of the vane need not involve the conscious coöperation of an observer identifying the vane during its motion, but the vane could be made a part of an automatic instrument, functioning in such a way as to give an indication on its dial only when the vane was subject to no force. Since "identifiability" in the fluid is needed only in order to provide an intermediate step in determining its velocity, and velocity is all that is needed in the equations in order to completely specify the situation, it would appear that in this way we have side-stepped the necessity for identifiability, at least in the fluid. This justifies our former statement that the demand for identifiability appears to be a "mental" or verbal demand, made in order to facilitate our thinking, and does not seem to be dictated by anything in the instrumental situation.

But although we can perhaps side-step the necessity for identifiability, there is one necessity that we cannot side-step, namely the necessity for a velocity. Whenever we have matter in flux we can analyze further and specify by some further physical operation (as by that of the forceless vane above) a further associated quantity, the velocity. We then can always define another quantity, the density, such that the product of velocity and density gives the total flux. At first the operation which gives "density" its meaning is a paper and pencil operation, the operation of dividing flux by velocity, and this sort of "density" would therefore be a pure convention. We know, however, that in a great many physical situations the density so determined may also be found by other sorts of physical operation. The discovery of these other operations is a further discovery about the physical nature of the system, and it is these other operations which give density its "physical" status as distinguished from a convention.

The velocity which we can associate in this way with the flux of matter is found in experience to have at least one other property, which is so universal that I believe that no physicist would ascribe "physical reality" to anything to which he ascribed velocity if it did not have this additional property. This additional property is associated with that change in our reference system which is the concern of special relativity theory. If we measure the velocity of some moving thing from a frame of reference moving with the same velocity as the thing (this means transporting ourselves, our instruments, our language, and our culture to a moving laboratory), then in this frame of reference the velocity of the thing will be zero. Or, "a moving object is reduced to rest in a set of axes moving with the velocity of the object." The necessity for this is obviously not contained in the operations by which velocity is defined, and probably the fact that such is always the case must be accepted as a further experimental discovery about the world. One must be prepared, however, to scrutinize each new case as it arises to find whether this association of properties continues to hold.

Let us now examine the flux of mechanical energy to see whether we can associate with it a velocity, and whether this velocity behaves in the way that we have come to demand in moving frames of reference. Consider our spring motor driving a flywheel through a belt from a frame of reference moving with the velocity of the taut half of the belt. We notice in the first place that there is a rather unexpected alteration in the path of energy flow. In the moving frame we still have to ascribe the same changes of energy as before to the motor and the flywheel, because our moving instruments still disclose a spring, in which there is a stress, unwinding, and a massive flywheel continually moving faster. But the taut half of the belt, which was previously the path of energy flow, has now become an object at rest in which there is a tension, and we have already seen that there can be no energy flux in stationary objects, no matter what their state of stress. The path of energy flux must therefore be somewhere else. It cannot be the slack half of the belt, because although this is moving with twice its original velocity, it still is in a condition of zero stress, and we can have no energy flux in moving objects without a stress. The only path remaining is the base; this moves from motor to flywheel with the same velocity which the belt had in the original frame, and it is the seat of a compressive stress. The conditions for energy flow are therefore found here in the base. Further, we now have the direction of the flux the same as the direction of motion of the matter in which the flux occurs, instead of opposed to it, as we found so paradoxically before. This reversal in the direction of flux is to be associated with the compressive stress which we now have instead of the former tension.

This switch in the path of energy flow brought about by a mere change in the frame of reference seems bad enough from the point of view of "physical reality" for the flow of energy, but there are other features still more upsetting. Consider the flux in the taut half of the belt in our original frame of reference. We disconcertingly found that a velocity had to be ascribed to this flux contrary to the velocity of the belt. We examine the assumptions back of this conclusion to see if we cannot find some loophole. We are pretty well compelled to assume that the path of energy flux is along the belt, because instruments outside the belt give no readings. Furthermore since the belt is linear, we do not see how any velocity which the flow might have can avoid being along the belt, and therefore either in the same direction as the belt or in the opposite direction.

The nature of the intellectual or verbal process involved in the thesis of the last sentence doubtless might be made the subject of an elaborate analysis, of unquestioned importance. Exactly what we have been doing here is not at all obvious; I think most people would be inclined to admit that whatever it is that they have been doing cannot be clothed satisfactorily in syllogistic form. Whatever it is that we do I would be inclined to characterize under the blanket term "verbal," perhaps not very helpfully to other people. The reason that I want to call it "verbal" is that I often catch myself in impulses which are obviously verbal: thus above, "If instruments outside the belt give no readings, then I have to *say* that the path of energy flow is in the belt." I try to check the consequences of "saying" that the energy flux is in the belt by visualizing what would be the physical consequences of its not being there, and I find my check in my inability to visualize such a situation. But I am uncomfortable if anyone should ask me how I know that my inability to visualize such a physical situation is not itself something which is primarily "verbal." The whole situation is obviously hazy; I may find that other persons will only partially accept the characterizations which I find valid, and it is even difficult for me myself to keep the meanings within the domain that I would like. But in spite of this, I suspect that everyone will find that he does, perhaps in other situations, use tools that he himself would be willing to describe as verbal in setting up his mental model. He says to himself, "Would I be willing to *say* such and such in such a situation?" and the model is so constructed that he can verbalize in congenial ways. The physicist justifies the construction of his model in this way because past experience has shown that it is profitable.

Going back to our example, in the original stationary frame we must "say" that the energy flux is along the belt. Furthermore, since mechanical energy is leaving the motor and accumulating in the wheel, we must say that the direction of flow is from motor to wheel, or paradoxically in the belt contrary to its motion. Or, in the language of "physical reality," "something" is moving in the belt with a velocity directed from motor to wheel. But now what happens in the frame of reference moving with the velocity of the belt? By the usual rule for the composition of velocities, whatever thing it was that was originally moving in the belt must now be moving from motor to wheel with a velocity about which we can say certainly that it is greater than the original velocity of the belt. But here appears a contradiction, because we have already seen that in the moving frame there is no flux of energy in the belt. There would seem to be no way of getting away from a contradiction here unless in the original frame the velocity of energy flow in the belt is the same and in the same direction as the velocity of the belt. Now surprisingly enough it turns out that precisely this is demanded by the mathematics. In the equations of elasticity theory a stress of tension and one of compression have opposite signs. When there is a simple tension the equations (A) indicate a velocity of flow equal to the velocity of the matter which carries the stress and in the *same* direction, but a *density* of energy which is *negative*. In this way the product of density and velocity gives a flux of the right amount and in the right direction. But what is the physical meaning of a negative density of energy? Any instrumental operations which could give direct meaning to a negative density of energy were so unsuspected from the original point of view that identified energy with ordinary matter that we simply failed to consider their possibility, and felt entirely secure in the verbal impulse which said that the energy in the belt *must* move in a direction opposed to the direction of the belt. However, the more detailed analysis justifies the mathematics. Our intuitive refusal of a negative energy density arose from an identification with ordinary matter. But we have seen that it is only *changes* of energy (divergence of flux) which have meaning in the equation of the first law. A negative density of energy may therefore be interpreted to mean that there is now inside a region less mechanical energy than there was at the epoch when we started our observations or calculations. Otherwise expressed, during the interval of observation energy has on the whole flowed out of the region. There is nothing strange about a situation of this sort, and we are reconciled. In fact, we take satisfaction that our mathematics, which has shown that we can meet the situation if we are willing to "say" that energy may have a negative density, has thereby proved itself capable of meeting better the verbal exigencies of the situation than our untutored instincts.

Nevertheless, I think it will arouse a certain dissatisfaction to realize that we have only been able to save the situation thus far by emphasizing the *difference* between energy and ordinary matter which permits us to give meaning to negative density of energy, whereas we started out with the resolution to completely identify the two. Our dissatisfaction is intensified when we analyze the general situation covered by the complete equations of elasticity. Consider the flux of energy in a right-angled wire, rigid, and carrying a tension in the two long members, *AB* and *CD*, and a corresponding shearing stress in the cross piece *BC*. (The conditions of equilibrium demand another small component of stress, which, however, may be made vanishingly small by making *AB* and *CD* indefinitely long.) Imagine the whole wire in uniform motion from right to left, as indicated by the arrow. Then the equations demand a flux of energy from left to right in *AB* and *CD* equal in magnitude to the product of tension and velocity. Continuity, as well as the equations, demands that there be a flux along *BC*. If it makes sense at all to talk about a "velocity" of this flow, then there must be a vertical component to this velocity in *BC*. (It would be interesting to know to what extent our conviction of the validity of this last statement rests on primarily verbal considerations.) Next, view the wire from a frame of reference moving with

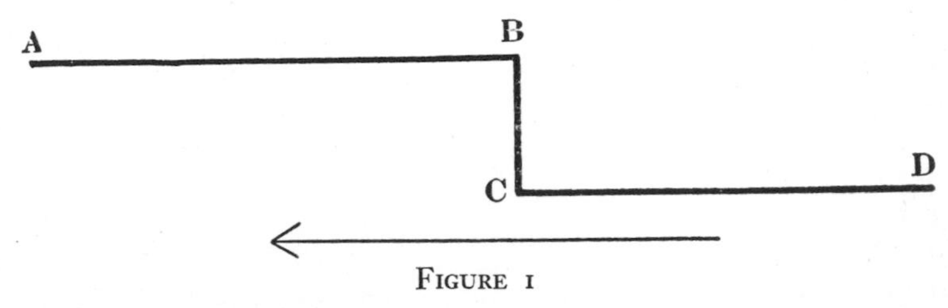

FIGURE 1

the velocity of the arrow. The wire is now reduced to rest, and hence there can be no energy flow in it, stress or no stress. In particular, there can be no energy flow in the arm *BC*, and hence any velocity of flow in *BC* is also zero, and all its components are zero. That is, we have reduced a velocity (the original vertical component) to zero by moving our frame of reference at right angles to the velocity. This it seems to me is irreconcilably paradoxical and utterly to be rejected.

In other words, the flux of mechanical energy cannot be completely paralleled to the flux of matter, because it is impossible in general to analyze the flux of energy into two factors, one of which has *all* the properties of velocity. Our urge to treat mechanical energy as a *thing* is a verbal urge, valuable within bounds, but capable of misleading and even false conclusions outside those bounds. One is inclined to moralize about verbalisms in general. Our instinctive verbalizing does not sufficiently realize the wealth of potentialities in experience. Because two branches of experience are similar in certain particulars we are inclined to jump to the conclusion that they are similar in all, and to cover them with a common verbalization. Thus above, we did not realize sufficiently that mechanical energy might have *some* of the properties of things without having them all. This verbal instinct is very deep-seated and it is active in the construction of all our language. But it must always be suspected, and checked by detailed and *exhaustive* inquiry when we are confronted with a new situation.

It is possible to restore verbal consistency to the general situation with regard to energy flow in stressed bodies, just as we have already seen we could restore verbal consistency in a simpler situation by accepting a negative density of energy, if we are willing to pay the price. A consistent verbalization is achieved by relativity theory; in fact one way of looking at relativity theory is that it is a method of achieving consistent verbalization subject to a few simple underlying physical postulates. The price that relativity theory demands that we pay for a consistent verbalization is that we give up thinking of energy as a thing with the simple properties of things and treat it instead as part of a higher complex, to the other members of which it is indissolubly tied. Consistency is achieved only by introducing the complete energy tensor of sixteen components, which are related not only to what we have been calling the energy, but also to the nine components of elastic stress, the three components of momentum, and the three components of the Poynting vector. All sixteen components must be considered in calculating the formulas by which any single component, and in particular the energy, is transformed to moving axes. This is not the place for further details, but it is to be recalled that we have to assimilate the flow of mass to the flow of ordinary energy, or otherwise expressed, to postulate the equivalence of mass and energy, and that the resolution of the infelicities which we encountered above in treating energy flow in a stressed right-angled lever involves slight variations in the masses, which may be accompanied by very large changes in the ordinary energy because of the astronomical factor of equivalence between mass and energy.

The magnitude of the intellectual feat of getting formal consistency into this complicated situation has justly aroused universal admiration. But without minimizing the intellectual accomplishment, it is not unfair to point out that no new physics can be created by an essentially verbal procedure, Eddington to the contrary notwithstanding. Also, because of the astronomical magnitude of the factor connecting mass and energy, it will be a long while before even the more obvious of the consequences of expanding energy into a tensor can be fully checked instrumentally. In fact, certain difficulties are already apparent, and there does not seem to be unanimity with regard to what to expect even among the relativists themselves, as is shown by a recent article of Eckart,* in which some of the difficulties of a relativistic treatment of heat and thermodynamics are suggested.

ANALYSIS OF FRICTIONAL GENERATION OF HEAT

Going back now to the first law, we have assumed thus far in the equation for the first law that we know what we mean by heat flow in the absence of any mechanical phenomena, and by the flux of mechanical energy in the ab-

* Carl Eckart, in *Physical Review*, 58: 919–924 (1940).

sence of thermal phenomena. We have recognized that there may be difficulties when heat flow and mechanical flux are not separated, but have hitherto side-stepped any detailed analysis of what may happen in such cases. It is instructive now to consider in detail a very simple example: a heavy block dragged by a string over a rough

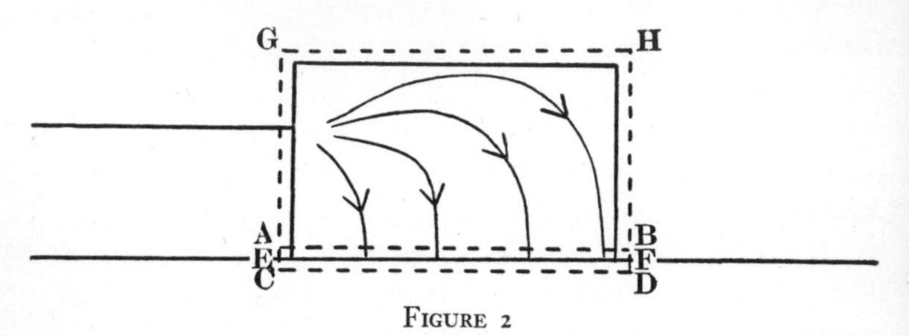

FIGURE 2

horizontal pavement, so that there is a constant generation of heat at the rubbing surface. How shall we describe the various fluxes in this situation? Mechanical energy flows into the block at its front end through the string by which it is pulled; there is a tension in the string and the direction of flow in it is therefore opposite to the motion of the string. It is this flow of mechanical energy in through the string which would ordinarily be described as the work done on the block. Inside the block there is a system of mechanical stresses, the detailed distribution of which would be complicated to calculate; in total these stresses are in equilibrium with the string because the block is not accelerated. The block, the seat of stresses, is in motion, so that in accordance with the general equations of elasticity which we have written down, mechanical energy flows inside the block. At the surface separating the block from the pavement, the stresses are shearing stresses to balance the tangential frictional drag; the equations indicate that here the path of energy flow is therefore perpendicular to the surface. The lines of flow of mechanical energy are roughly as indicated by the lines in the figure. In the pavement immediately under the block, there is an equal and opposite set of shearing stresses, and these transform themselves into other types of stress as we penetrate from the surface into the body of the pavement in a way which is determined in detail both by the way in which the pavement is held stationary and by its elastic constants. The details of the stress distribution here are of no immediate moment, however, for since the pavement is stationary there is no flux of mechanical energy in it. The lines of flow of mechanical energy therefore terminate abruptly at the surface separating block and pavement. If we draw two surfaces close together, one in the block and parallel to its surface, and the other parallel to it in the pavement, as indicated by *AB* and *CD* in the figure, then we have a finite amount of mechanical energy flowing into a region of infinitesimal volume. This ought to produce some infinite effect unless there is some compensating action. Of course we know what happens; the mechanical energy which flows with a bang into the surface of separation is converted at the surface into heat in equivalent amount, and this heat flows out of the infinitesimal region through the surfaces *AB* and *CD*. This heat flow from the surface into block and pavement may be measured instrumentally by determining the temperature gradient normal to the surface with miniature probing thermometers, and combining with the thermal conductivity of block and pavement. We shall assume that these are of the same material and that the other relations are symmetrical, so that half the heat generated at the surface flows into the block and half into the pavement.

How now shall we go about applying the first law to this situation? We must first mark out the region to which we are to apply the law; we then post sentries at the boundary who measure the mechanical energy and the heat entering the region. First apply the law to the region indicated by *AGHBA*. Mechanical energy enters at the front end through the string, and an equal amount flows out on the elastic Poynting vector at the surface *AB*, so that the total *W* for this region is zero. Heat enters the region across the surface *AB* in amount equal to one half the mechanical equivalent of the work done by the string. The effect of this net inflow of heat is to increase the internal energy by raising the temperature of the block at a rate which we could at once calculate if we knew its heat capacity. We are not disturbed to anticipate the meaning of "internal energy" in this simple case.

Next let us take the region *CGHDC* as the region to which to apply the first law. Mechanical energy now flows in on the string, and there is no outflow, so that *W* for this region is the total work done by the string. Heat, on the other hand, flows *out* of this region in amount equal to one half the equivalent work done by the string, so that the net input, *W* plus *Q*, for the region *CGHDC* is the same as for the region *AGHBA*. It follows that the rate of rise of temperature of the region *CGHDC*, which has the same mass as the region *AGHBA*, is the same as that which we just calculated for the region *AGHBA*. This, of course, is all as it should be.

What happens now if we delimit the region by the surface of discontinuity itself, *EF?* Difficulties at once present themselves. How are we to visualize this surface physically? Is it drawn in a vacuum between block and pavement, and does every atom of the block (how easy it is to talk of the atom!) forever remain on one side of the surface and every atom of the pavement on the other? If so, in what does the mechanism of friction consist, and how does the pavement exert a force on the block? We have to deal with the situation verbally. It turns out that we can invent a consistent verbalization, but I very much question whether we could do this if we could not see in advance the answer we want. If we want the surface of separation to be free from the excursions of atoms back and forth across it, then we must "say" that the atoms of the block are subjected to the action of forces exerted "at a distance" by the atoms of the pavement. As the block moves along the pavement, one end of the line of action of these forces is continually being displaced, so that work is continually being done by the atoms of the block. On the other hand, the atoms of the pavement are stationary, so that at their end of the fictitious lines of force no work is done or received. We have the paradoxical result that the work received by the block from the pavement across the surface of separation is not equal to the work done on the pavement by the block. This failure of equality of direct and reaction work always occurs at a surface where there is tangential slip and there are forces in the direction of slip. This sort of thing does not occur very often in the conventional thermodynamic analysis; I think that many physicists have a sort of instinctive feeling that direct and reaction work are always equal. Obviously this can be the case only when there is no discontinuity in the motion at the surface, as at a piston which is compressing a gas.

It is instructive to extend the same method of treating the mechanical action at the surface of contact of two bodies to other examples, such as that of an automobile traveling at constant speed under the action of its engine, assuming that all the retarding friction is rolling friction between the wheels and the road, or the same automobile, with its engine cut off, coming to rest under the action of its brakes. We shall not go through the detailed analysis, but anyone familiar with the problems of elementary mechanics knows that a mathematically satisfactory discussion can be carried through on the basis suggested. I think, however, that if one stands back and looks at what he has done he cannot help wondering a little at the lengths to which he has been driven. The instruments and the concepts of elementary mechanics are large-scale affairs, but here we find ourselves almost forced to talk about "atoms" and "forces at a distance" between the atoms. We have gone far beyond the spirit of our original laboratory problem.

What now does the first law say about the block, delimited by the surface *EF*? We must first make the situation of the surface *EF* still more precise — let us assume that it is exactly half way between the two surfaces, cutting in two the imaginary lines of force between the atoms in surface of block and pavement. We can now "say" that the displacement of the mid-point of the line of force is only one half the displacement of the terminal atoms in the block, and that therefore the flux of mechanical energy across the mid surface is only one half the flux out at the surface atoms, which in turn is equal to the flux in on the string. On the other hand, we must "say" that there is no flux of heat across the mid surface, either conduction or radiant heat, merely because of considerations of symmetry. This leaves the net flux of energy into the block one half the work done by the string, as before. We have therefore achieved a consistent verbalization, but at the expense of what strangeness in the instrumental operations! For we talk about a mechanical work at the mid point of a purely imaginary line of force in empty space acting on nothing, and to be consistent we would have to talk of a conversion of mechanical energy into heat in the empty space of the upper half of the imaginary lines of force. Surely in the imagined empty space between the two surfaces our sentry would find nothing which he could instrumentalize into either a flow of mechanical energy or of heat. We could not have achieved our consistent verbalization if we had not known the answer.

It follows that we must not carry over the verbalization into new situations in which we do not already know the answer. It is not hard to see where the difficulty originates. The dividing surface is a surface of mathematical discontinuity, a purely paper and pencil thing for which we never find a complete physical counterpart. The instrumental operations for determining heat flow demand the determination of a temperature gradient, and this demands the instrumental determination of temperature at two different points, and the instrumental performance of something that corresponds to taking a limit. But if we have a surface of discontinuity always present, we can never locate either of the points which give us the gradient in the surface itself, and hence can never get the gradient *at* the surface. The mathematical conditions are inconsistent with the physical prerequisites; there are two limiting processes involved, one in the concept of surface of discontinuity, and the other in the concept of gradient, and we are required by the mathematics to perform these two limiting processes in an order which the physical situation does not allow.

If we are willing to commit ourselves to the atomic and kinetic picture of what happens at the surface of contact of block and pavement we are not disturbed by these various difficulties, for we have the two sorts of atoms promiscuously passing back and forth across the surface, colliding with each other, and continually initiating irregular mechanical action which eventually degenerates into heat as the scale of the action becomes finer. Or if we are troubled by being forced to an atomic mechanism for large-scale phenomena, we can imagine a thin liquid film between block and pavement, the liquid being in smooth laminar motion everywhere parallel to the surfaces, and in virtue of its viscous resistance everywhere the seat of conversion of mechanical energy into heat. So long as the thickness of the film is finite, we can draw the separating surface anywhere in the liquid and give a perfectly consistent and coherent account of the flow of mechanical energy and of heat into block and pavement above and below the surface. There is now no difficulty in letting the thickness of the film become vanishingly small, and thus getting a valid physical picture without getting away from the macroscopic representation of matter as something continuous and capable of infinite subdivision.

This, however, is more or less by the way. A more interesting point is that we would do well not to attempt to apply the conventional statement of the first law to a region whose bounding surface is a surface of mathematical discontinuity where mechanical energy is converted at a finite rate into thermal energy. Our paper and pencil manipulations may make us want for other reasons to set up the fiction of such a surface, but we must be prepared to find that the instrumental operations which give simultaneous meaning to $\vec{w}$ and $\vec{q}$ fail. In other words, our verbalizing (that is, mathematicizing) may lead us into situations in which we cannot assign a meaning simultaneously to flux of mechanical energy and flow of heat, contrary to the tacit assumption in the conventional statement of the first law. In the example considered the difficulty has not been serious, because we have been able to side-step it by drawing the dividing surface either just inside the surface of the block or of the pavement.

The situation under other circumstances may become much more serious. The difficulties may arise not from our verbal impulses, but from the physical situation itself. This may be so complicated, as in the case of small-scale turbulent motion in a liquid, that we would not know what instrumental operations to perform to separate the $\vec{w}$ from the $\vec{q}$. The difficulties appear to be particularly formidable in the case of radiant energy. The general conclusion is plain enough: the first law in the form

$$\text{(time rate of change of energy)} = -\text{Div}\,\vec{w} - \text{Div}\,\vec{q}$$

may not always be applicable, for $\vec{w}$ and $\vec{q}$ may not always have a meaning.

Not only must $\vec{w}$ and $\vec{q}$ have a meaning for the sentries at the boundary of the region, but, more generally, we have yet to discuss what to do if anything happens to them which they cannot report either as a flux of heat or of mechanical energy. In particular, we have not yet allowed any flow of "matter" across the boundary. If the original surface is drawn entirely inside stationary matter, the $\vec{w}$ and $\vec{q}$ and $d\Delta E/dt$ of the first law refer to the matter inside this surface, or to a fixed "body." In the usual applications this restriction to a definite and fixed "body" or piece of stationary matter is tacitly understood, and I shall understand this restriction for the discussion in the next paragraph. There is no necessity for such a restriction, however, for the region to which the law is applied might equally well be a region of "empty" space, "occupied" only by an electromagnetic field (that is, instruments inside the region disclose the presence of such a field), and the surface bounding the region may be in such a state of motion with distortion that the total volume inside the surface is changing. In this case, there will be a contribution to the flux of mechanical energy across the surface arising from the motion of the surface through a region in which there is localized electromagnetic energy, and this flux again will be specifiable in terms of instrumental operations performed by the sentries stationed at the surface and moving with it. In this case of a moving surface in an electromagnetic field nothing happens to the sentries which they cannot describe as flow of heat or of mechanical energy, and the first law applies at once. But the sentries cannot deal with the situation without further instructions when they are moving through "matter."

THE MEANING OF ΔE

Assume now that we have a region for which $\vec{w}$ and $\vec{q}$ have meaning for the sentries posted at the boundary, and furthermore that the boundary marks off a definite stationary "body." Let us examine the meaning of the ΔE. The ΔE refers to the region inside the boundary, and under these conditions is usually called "the increase of internal energy of the body." "Internal energy," as it is commonly used in elementary thermodynamics, does not include such things as kinetic energy of molar motion of the parts of the system, or their potential energy. The reason for this is merely that elementary thermodynamics is seldom concerned with transformations involving large-scale kinetic energy. But obviously such effects are to be included, and whenever in the future we talk about the "internal energy" or the "energy function" or merely the "energy" of a body, kinetic and potential and similar terms are understood to be included.

A naïve meaning for the law as formulated in the equation would be that the increase of internal energy of the body is to be determined by some sort of independent operation on the body, and that then it will be found as a matter of experiment that when the sentries report the entry into the region of a certain amount of work and heat, the operation on the body inside will give a result which is just equal to their sum. This naïve statement, however, requires considerable qualification. There is no simple instrument which, when applied to a body, will give a result such as to satisfy this condition; neither is there any combination of instrumental and paper and pencil operations which can be formulated without reference to the law itself which will give such a result. This does not mean, however, that the first law is a tautology, and that the ΔE which is the sum of W and Q is the sum by definition only.

For the purposes of this immediate further discussion it

is convenient to write the law in the conventional differential form

$$dE = dW + dQ.$$

If the changes for any actual body are not small and the material of the body is not under homogeneous conditions, we assume that we can split the body or the process into parts small enough so that the differential form will apply. The essence of the first law is now contained in the mathematically different character of dE on the one hand and dW and dQ on the other; dE is a perfect differential, as distinguished from dQ and dW. Strictly, these two sorts of small quantities should not be written in the same way, and some careful authors do actually use a different notation for them. The statement that dE is a perfect differential implies that there are certain variables of "state" of the body such that when these are brought from fixed initial to fixed final values the sum of the heat and the work for the change is always the same, whatever the precise method of change. Or, expressed in another way, whenever the body (or the interior of the region) is brought back to its initial "state" the sum of heat and work absorbed during the complete cycle of change will be found to be zero. It is the statement that dE is a perfect differential in the parameters of state that removes from dE the character of a pure convention, defined by the equation. Under these conditions the equation formulating the first law really says something.

What is to be understood by "state" of the body? There is danger that it may degenerate into a tautology, because we might say that by definition the body has returned to its initial state when the total work and heat of the sequence of processes is zero. There must be more physical content to the concept of state than this. Actually, as already suggested in the discussion of temperature, "state" has an independent significance, and connects with the "properties" of the body — a body has resumed its initial state when all its properties have resumed their initial values. The first law states that there is some function of the parameters that determine the state (or the properties) of the body such that the difference of the function for two different states is the sum of the net heat and the work entering the body during the change from one state to the other, no matter what the details by which heat and work are imparted to bring about the change of state. This involves the experimental fact that there is an infinite variety possible in the details by which a given change of state may be brought about.

If at first we are unsuccessful in finding a set of parameters which determine an energy function, we do not rush to conclude that the laws of thermodynamics do not apply to our system, but rather suspect that we have not a complete list of parameters of state. In particular, we may suspect that there is microscopic turbulence on a scale too small to deal with with instruments of our present scale. If this is the true solution, then experience suggests that if we wait a little longer to make our measurements, the turbulence will die out, as it always does when given time, and that then our thermodynamic description will become adequate. If, after waiting, our measurements still do not give a single-valued function for E in terms of pressure and temperature, for example, then we may suspect that other *sorts* of parameter, such perhaps as a magnetic field, are involved, and we will institute a search for physical phenomena previously overlooked or dismissed as unimportant which may give the clue to the nature of such additional parameters.

Are we not asking too much when we demand that we be sure in any practical case that we are in possession of *all* the parameters required to fix the state of the body? Is this not equivalent to demanding that we know *all* the properties of the body, or at least all its independent properties? But when magnetic phenomena were still undiscovered, no one suspected any such property of a body as its magnetic permeability, and how can we be sure that some analogous phenomenon may not be awaiting us now? I think we can never have such an assurance. We get a glimpse here of a sort of super-principle that should control the formulation of any scientific theory — every theory should be formulated in such terms as to leave room for the future discovery of kinds of fact at present unknown. If a theory is not so formulated it loses part of its aspect as a program of action, which makes theories so valuable, and becomes a description of history.

The first law, then, can merely state that there is an energy function in terms of the parameters which we now know. Or, up to the present (1941) no known case has arisen in which an energy function cannot be found in terms of instrumental operations which we can now perform. If at some time in the future we discover how to perform instrumental operations of which we now have no inkling and thus find new properties and thus new parameters of state for familiar bodies, we know from the present validity of the first law in our present restricted universe of operations that there can be no cross connections between the new undiscovered parameters and our present ones. That is, by changing the present parameters we cannot bring about changes in the undiscovered properties which are associated with energy changes. Our present scheme of energy changes is complete; the hypothetical new properties cannot overlap.

Since the prior evaluation of an adequate set of state parameters is fundamental to all thermodynamics, it will pay to stop for a little further consideration of the criteria which we apply when we decide in any special case that the parameters which we now know are adequate for our present purposes. It seems to me that we are satisfied that our characterization of a body is complete if, whenever in the future the body again is characterized in the same way, every feature of its behavior again is the same. "Every feature of its behavior" means the result of every operation that we can now perform on it and its future course of action in any specified environment. This criterion patently assumes that we are dealing with "causal" systems, that is, systems in which the future behavior is connected by regularities which can be formulated with present condition and past behavior. It is a matter of experiment that most of the objects of daily life are of this sort that it is possible to perform operations on them such that their future behavior in any environment can be completely predicted in terms of the results of these operations. Without the existence of such objects science in general and thermodynamics in particular would not be possible. Nevertheless, there are objects of not too frequent occurrence which do not obviously have this characteristic. Such objects, for example, are any exhibiting hysteresis. A piece of iron which has a complicated magnetic history may respond in a great variety of ways to alterations of the magnetic field to which it is exposed, even if its magnetic field and induction, the parameters which are usually considered to determine the magnetic state of a body, have assigned values. The existence of such bodies does not necessarily mean that we are dealing with something "non-causal," for no one would maintain that the future behavior of a piece of iron showing magnetic hysteresis is not completely determined by its *entire* past history. Nevertheless, the existence of such bodies is embarrassing, because, as commonly used, the notion of "state" and "property" involves only the results of operations which can be performed on the body *now*, and does not involve a knowledge of past history. A "property" of the body is such a thing that we are able to say of it the body "has" such and such a property *now*. We probably would not say of a certain piece of iron that it was one of its properties to have been red hot at six o'clock last night. As always, the distinction between operations performed now and operations performed in the past cannot be made precise, and there may be subtle questions as, for instance: is the velocity of a body one of its "properties," when we consider that to determine its velocity we have to find its position at *two* different times, one of which must obviously be in the past? We probably would be satisfied with a "now" which had as much indefinite extension as the psychological "now." This would permit us to make instruments by which the velocity of a body would be indicated by a pointer on a scale, perhaps by some governor mechanism, as in some types of automobile speedometers. By the same token, we can construct instruments such that the acceleration of a body is indicated by the position of a pointer on a scale, and we call the acceleration one of the "properties." I think it is not necessary to be too meticulous, and that we can be agreed in most of the cases of practical interest whether to call the operations present operations, and whether we have determined a "property" in applying them. In particular, the elementary thermodynamic properties of a body are such that we can say of them "the body *now* has these properties."

In order to bring bodies which exhibit hysteresis under the thermodynamic notion of states, an excursion has to be made into the paper and pencil domain. In the case of the magnetic body the thesis is that there is a fine structure which is not disclosed by those large-scale measuring operations which give the ordinary B and H. The complete parameters of state would thus demand a characterization of all the individual elementary magnetic domains. The instruments necessary to make this characterization would obviously be small, but they still would be far above the molecular range, and the parameters determined with them would still be the macroscopic parameters assumed in thermodynamics. The actual experimental justification of this thesis for every body showing any kind of hysteresis would obviously be difficult and certainly has not yet been carried through, but we appear to be led into no inconsistencies by such a thesis, and we are often led into suggestive experimental programs. The energy concept may in many cases be applied to bodies showing hysteresis by reducing them to some standard state which does not show hysteresis, as by dissolving a magnet in acid. There are also cases in which the notion of "states" becomes applicable, and hysteresis is removed, by the introduction of a single new parameter, as for example, certain classes of alloys become entirely manageable when a single new parameter measuring molecular "disorder" is introduced. Some general method would be desirable of finding whether one or more new parameters would effect the desired reduction to "states."

There are other kinds of objects for which it is not at all evident that the causal character could be saved by operations this side of the molecular. An example is probably the capricious freezing of a sub-cooled liquid, and certainly an example is an atomic disintegration. Such objects are not handled by the conventional thermodynamics: whether they could be handled by some extension in the spirit of thermodynamics is a topic for further consideration.

In this connection one cannot, of course, avoid thinking of quantum mechanics, with its thesis that all small-scale happenings can be handled only statistically. One can adopt at least two points of view here. One could maintain that *all* the parameters had not yet been determined for small objects, because by definition all the parameters are not known until we have a causal system. Very much this attitude is adopted by many physicists, particularly those of the older school. But in the face of facts as unlike those presupposed in thermodynamics as these, it seems to me that such an attitude becomes predominantly verbal and not sufficiently profitable. What the quantum physicist does is to change his definition of a "complete" set of parameters. For him a "complete" set of parameters is the one which we get after performing all the operations of which we are capable — "we do our darndest — angels can't do more." If such parameters do not constitute a causal set, we just have to make the best of it. Experimentally, we find that many systems exhibit *statistical* regularities when defined with such parameters, and this can be made the basis of the only even approximately adequate theory of such phenomena that has yet been proposed. The justification for the thermodynamic approach is that there is a very large group of phenomena which are adequately treated in terms of macroscopic operations. And if one is inclined to ascribe a greater "fundamentality" to the microscopic statistical approach, it should not be forgotten that all the basic microscopic operations are ultimately defined in terms of the macroscopic operations of thermodynamics. Logically, so far as the situation can be reduced to logic, the thermodynamic approach seems the fundamental one.

The first law is established by the possibility of actual construction of the energy function in every specific case. Apart from the first law itself, there is no general rule by which an energy function can be constructed (each new material requires separate experimental treatment); there is no instrumental operation or paper and pencil operation on the data of the instrument that gives the energy function apart from the operations implicit to the formulation of the first law.

Given any new body, the construction of its energy function demands in the first place an exhaustive study of its properties and the parameters on which these properties depend, in order that the "state" of the body may be specified. This preliminary study is not thermodynamic in character. The difference of energy between any two states is then to be found from the sum of the heat and the work required to bring the body by any convenient process from the one state to the other. In the detailed working out of the energy function for different bodies there are various simplifications that are too well understood to elaborate in detail here. Perhaps the most important of these is that we do not need to work out the energy function for every one of the infinite bodies which we meet, but that we are concerned only with the "kind" of body or material. One pound of iron has the same energy function as any other pound of iron, and in general all the thermodynamic properties are the same. (If *all* the properties were the same we could not speak of two *different* pounds of iron.) Again, the energy function for any piece of iron weighing ten pounds is ten times the function for any piece weighing one pound.

LIMITATIONS ON THE GENERALITY OF ΔE

How general is the energy function which we have thus found how to construct? It obviously has great generality. If we have once found by the method outlined the difference of energy between two states of the body, then we know the sum of the heat and the work required to bring it from the initial to the final state not only under the original conditions as part of some definite physical system in some definite enclosure, but also when it is a part of other physical systems, in different sorts of enclosure, and coupled in different ways to other sorts of body. But there is nevertheless a limitation on this generality, because the detailed construction of the energy function has demanded that we get the body from the initial to the final state by some one process for which heat and work are defined. Having once found such a process and performed it, we may then forget it and still have the same difference of energy between the initial and the final state even if the passage between these two states is brought about by a process for which heat and work are not defined. In this way we attain a degree of independence of our initial restrictions on heat and work.

This is a matter of the greatest importance, for, as already intimated, it is only under highly exceptional conditions in practice that a clean-cut analysis can be made into heat flux and flux of mechanical energy. Consider, for example, Joule's original set-up for determining the mechanical equivalent of heat by the elevation of temperature of a pail of water stirred by paddles. The conversion of mechanical energy of the paddles into thermal energy of the water is to a certain extent a degradation phenomenon, involving the transformation of large-scale motion into turbulence of an ultimately molecular scale. One of our sentries, posted at an element of the boundary of the liquid, would be hard put to know how to analyze what he observed into heat or mechanical work, and his result would certainly depend on the scale of his measuring instruments. But although we cannot follow the details of the process by means of a first law formulated in terms of heat and mechanical work, the energy concept does make us master of the broad general features. For the total mass of liquid may be split into elements so small that each of them may be treated as homogeneous and therefore as having a known energy function, and then by integration of the energy increments for all the elements, including in the energy of each element its kinetic energy, we obtain an equality with the total heat put into the whole large-scale system, the boundaries of this system being drawn at such a distance that heat and work are there defined, as can obviously be done.

In particular, if the boundary of the whole system is such that no heat or work crosses it, then the total energy of the material inside is constant, no matter whether the processes which occur within are so complicated that heat and work have no meaning in detail. This statement that the total energy within is constant is susceptible of experimental verification because the energy of each element is a known function of its parameters of state. A special case arises when we have made measurements of the energy of only part of the material within the enclosure; the first law then enables us to predict what we will find for the energy of the rest of the material when we make the measurements on it.

A study of the most general conditions under which heat flow and work have meaning would doubtless be of great interest, and so far as I am aware has never been attempted. However, it is pretty obvious what a sufficient set of conditions is. If the process is "reversible," so that an infinitesimal change in the conditions reverses the direction of the process, which has as a corollary the consequence that the process is being performed with infinite slowness, I think it fair enough to suppose that an analysis into heat and work would not cause our sentries any insuperable difficulty. Conventional thermodynamics is, of course, built on the reversible process, and one reason for it probably lies here. However, infinitely slow processes are not exclusively necessary, for obviously examples can be given of purely mechanical processes running at a finite rate, as the speeding up of a flywheel by pulling off a string wound around an axle. Neither should a mere transfer of heat at a finite rate make any difficulty, as when heat flows down a copper bar whose ends are at 0° and 100°. But if there is *conversion* at a finite rate, I believe there will always be ambiguity, as exemplified by the degeneration of turbulence into heat in a viscous liquid.

Given now a substance such that it is possible to find the conditions by which it may be brought from one state to another in such a way that heat and work are defined for the process, then its energy function may be determined as a function of the parameters of state. For such a substance, once the energy function has been determined, the first law in the form $dE = dW + dQ$ has its naïve significance. For if any process is performed for which dW and dQ are defined, the sentries at the boundary may make independent determinations of dW and dQ, and at the same time the observer in the interior may make an independent determination of the dE of the matter inside, and then it may be checked that the results of the different observers are connected as demanded by the equation. The truth of this statement is unaffected by the fact that the operations by which the interior observer determines dE are operations peculiar to the specific substance, and involve a combination of operations with instruments and with paper and pencil, and that the original discovery of these operations was made under conditions such that dW and dQ were defined.

METHODS OF EXTENDING THE GENERALITY OF ENERGY

We have already seen that the energy function, once having been found, can be used to determine the behavior under conditions broader than those originally contemplated, namely under conditions for which dW and dQ are not defined. This fact can be used to dispense with the necessity, in originally evaluating the energy function, for always designing the conditions so that dW and dQ have meaning. Suppose that we have two different pieces of matter. One of them we suppose to have been already well explored and its energy function to have been originally determined under canonical conditions for which dW and dQ were defined. The other of them we suppose to be of some unknown new material whose energy function is not known and which we are not even sure can be made to take part in processes with defined dW and dQ. These two pieces of different sorts of matter are initially given us, each in a homogeneous condition, but not necessarily at the same temperature and pressure. We now bring these two pieces together into an enclosure bounded by a surface impervious to the passage of heat or of mechanical energy. The two pieces of matter are allowed to react with each other for an interval within the enclosure. It is indifferent whether the interchange can be expressed in terms of work and heat or not, and during the process of interchange the pieces of matter need no longer be homogeneous. After the reaction has run to any desired extent, the two pieces are separated, still within the enclosure, and held isolated from each other for a sufficient length of time for each to resume again a homogeneous condition. We assume that we know what the parameters of state of the two pieces of matter are, and that we determine the values of initial and final parameters of both. Then, because we know the energy function of the first piece of matter, we know the change of energy which it has experienced during its reaction with the other. This must be the negative of the change of energy of the second piece, since the total change of energy of the two pieces together is zero. That is, we have a method for finding the changes of energy of the second piece without exposing it to conditions such that dW and dQ have meaning. By sufficiently varying the conditions of reaction we can completely map out the energy function of the second piece of matter.

There are cases in which some such generalized procedure for evaluating the energy function seems to be absolutely necessary, that is, pairing off the substance whose function is to be evaluated against some other substance whose function has already been determined under canonical conditions. An example would be to find the difference of energy between a diamond and the piece of graphite into which it may be transformed by heating it to a high enough temperature. There is no known method by which graphite may be transformed to diamond, and no known method by which diamond may be transformed to graphite under conditions such that dW and dQ have meaning. (Notice parenthetically that this latter statement is very nearly equivalent to saying that there is no known way of transforming diamond to graphite reversibly.) However, diamond may be transformed to graphite by heating in an adiabatic calorimeter, and we do not need to elaborate in detail to show how the energy change is found from the known change in the energy of the other contents of the calorimeter during the transformation.

The concept of internal energy as generalized thus far applies strictly only to pieces of matter in a homogeneous condition. The concept is at once extended, as already suggested, to matter which is not homogeneous. This is done very simply by cutting up the finite non-homogeneous piece of matter into infinitesimal elements so small that each may be regarded as homogeneous, characterized with sufficient accuracy by the average values of its parameters of state. The energy of the whole piece is then defined to be the sum of the energy of its elements. This is not a tautological definition, for in manipulating the energy thus defined further experimental knowledge is assumed in postulating that the total energy of non-homogeneous pieces of matter reacting inside impervious enclosures is constant.

In extending our range to include non-homogeneous matter we have to reëxamine some of our fundamental procedures. For example, if parts of the system are in relative motion we have to reëxamine our whole procedure for defining and measuring temperature. We find, for instance, that a thermometer with flattened bulb gives different indications in a stream of water according as it is placed broadside on or edgewise in the stream. Facts like this lead to the eventual conclusion that the only "proper" way of measuring the temperature of a moving object is with a thermometer moving with the object. In fact, our fundamental fact that systems come to temperature equilibrium applies only to systems at rest, so that in particular the thermometer must be at rest with regard to the system whose temperature it measures. We read the moving thermometer as it moves past us, or if we use a minute thermocouple we read it on a stationary galvanometer, twisting the leads so as to introduce no error from their motion. We explicitly disregard relativity effects here. Similarly the instrument with which the other parameters of the elements are measured, such as the pressure gauge, must be moving with the parts of the system.

At first, the temperature which we assign to a small element in a non-homogeneous medium — whether the inhomogeneity consists in relative motion or other non-uniformity — by procedures with small relatively stationary thermometers is to be regarded as a purely conventionalized definition of temperature under such conditions. Such a convention has an inevitable naturalness, because this temperature approaches the originally defined temperature when the body approaches uniformity. The profitableness of such an extension comes from the great range of the manipulations into which the extended concept fits: for instance, the fact already mentioned that the energy of a system is the sum of the energy of the elements each regarded as homogeneous, or the fact that it is possible to calculate such things as the way in which the indications of a thermometer with flattened bulb changes as its orientation to the stream changes, or that given the inhomogeneity we could calculate an upper limit for the size of a thermometer to give satisfactory readings, or, conversely, given the thermometer, we could calculate a degree of inhomogeneity that would make its readings unsatisfactory. In fact, it is only the possibility of this sort of thing that eventually justifies thermodynamics at all. For whenever there is mechanical work there is relative motion: the justification of applying the temperature concept at all in such systems can be afforded only by a numerical computation that shows that in the limit as the motion is made slow the neglected effects really do vanish in comparison with those retained.

There is no question, I think, that the physicist does, with complete confidence, generalize the energy concept in the way suggested so as to apply to non-homogeneous matter. The direct experimental justification, however, I suspect to be very meager. The reason is doubtless partly that the experiment would be very difficult, and partly also that the question is not often raised. In fact I suspect that the average physicist is so confident of his position here that he would have little patience with anyone who would "waste his time" in attempting seriously an experimental verification. Part of his impatience doubtless arises because he can see a loophole which would enable him to "explain" any apparent failure of the experimental verification. If the energy of the elements, regarded as homogeneous and like bulk matter, did not add up to a constant, he would simply say that other parameters are required to completely fix the state of the elements in addition to those which suffice for large homogeneous pieces of matter. For instance, if the total energy of a piece of copper in which there is thermal conduction down a temperature gradient does not add up to that of the elements, to each of which is assigned the internal energy obtained from measurements on pieces of copper at uniform temperature, then there is an obvious way out in saying that the internal energy of a piece of copper depends not only on its temperature but also on its temperature gradient. That is, the parameters of state are temperature and temperature gradient instead of merely temperature. It is easy enough to verbalize in this way, but the instrumental processes by which operational meaning can be given to such verbalizations may be complicated, and furthermore may demand a prohibitive amount of experimental accuracy in view of the fact that the original failure of the energy to be additive was doubtless small. The physicist is therefore almost certain to rest at the verbalization and let it go at that, making perhaps the further observation that he has no positive knowledge of any phenomenon which indicates that he could not carry his program through if he had to. This attitude of the physicist, which might strike one as unpleasantly sloppy, has after all its justification. As a matter of fact, the physicist never is concerned with measurements on systems which are too inhomogeneous. Instrumental exigencies, such as the scale of his instruments and the time required for the instruments to come to a reading, demand that the subjects of his detailed study be nearly homogeneous and nearly in a condition of equilibrium. His measurements are made at points of rest or nearly of rest, and all his experience indicates that the energy concept is adequate under such conditions. If he wants to push the energy concept far beyond these points of rest, that is largely a verbal matter to which he is impelled only because his paper and pencil operations have no mechanism for cutting themselves off automatically, like his instrumental operations, when he has penetrated to a certain scale of magnitude.

The justification and convenience of the verbal or paper and pencil extension of the energy concept to which the physicist is thus naturally led is that after burrowing around awhile in the submerged paper and pencil world he often is able to emerge again into the actual world of systems approximately at rest and of the scale of feasible instrumental operations, and whenever he does so emerge he finds that his paper and pencil energy concept has guided him to the right point. By far the most important verbal or paper and pencil extension of the energy concept is into the atomic or sub-atomic region of statistical mechanics. Of the fruitfulness of this extension there can be no question; my concern here is that we should not forget what it is that we have really been doing.

It appears that the "energy" which occurs in Schroedinger's wave mechanics equation is not so immediately connected with the operations of the laboratory as our generalized thermodynamic "energy," and for that reason is to be distinguished from it. The energy of Schroedinger's equation is something of which we can say: "It is possible by the methods of trial and error to associate with every system such a number E that all future changes in the system which take place within an impervious envelope can be described with wave equations with a constant E." The wave mechanics energy is therefore still more an affair of paper and pencil than the energy of thermodynamics.

It is often said that energy is only a constant of integration. It is evident that if a system is controlled by differential equations, there will always be constants of integration, determined by the initial conditions. Or still more generally, even if the differential equations do not exist, but the future development of the system is determined by its past so that it is possible to reconstruct the past from its present condition, then there must be constants pertaining to the initial configuration recoverable by operations performable on the system at any stage of its development. These operations, of course, include pencil and paper operations. Although these considerations demand the existence of certain constants associated with any system, they do not determine the exact operations by which the constants are to be found, and it is further in the exact nature of the operations for determining energy that the particular properties of energy are to be found. One could not predict from a knowledge of the mere existence of a differential equation controlling the motion of a simple mechanical system that the sum of the squares of the velocities of each element into the corresponding masses would be constant. That precisely this is constant is determined by the specific form of the equations of motion. The exact operations by which energy is determined in general are fixed by the specific form of the underlying equations, or more generally by the specific form of the rule by which the past may be reconstructed from the present.

As the physicist has come to use it, the energy concept is generalized in still another way beyond the operations of the original definition. These operations demanded the separation of the universe into at least two parts: an inside part, with the changes of energy of which we are concerned, and an outside part. Furthermore, an observer external to the inside part is demanded who makes instrumental and other operations on the part inside, and thus evaluates a change of energy which he uses for his own purposes. From this point of view it is therefore completely meaningless to attempt to talk about the energy of the "entire universe," which by definition can have no parts outside it and no external observer. (Notice parenthetically that the observer need not be "external," in the geometrical sense, to the system whose energy is being determined, but his externality may consist merely in that his operations of observation and measurement are entirely without reaction on the system.) There is no question, however, but that the physicist has generalized the energy concept in such a way as to permit him to talk about the energy of the entire universe, or more properly the difference of energy of the universe in two different states. Furthermore, he has a perfectly clean-cut operational meaning for this extension. He would find the difference of energy of the universe in two configurations by dividing the universe into enough small pieces so that each could be regarded as homogeneous, and then adding the changes of energy of each of these pieces. The change of energy of each of these homogeneous pieces would have to be determined by the methods which we have already discussed, which in the first instance demand isolation of that piece. He would also recognize, of course, that all the energy of the universe is not localized in the matter which composes it, but that a large part is associated with the radiational field, and that to catch this part of the energy, observations are necessary in the "empty" space surrounding the material universe at distances which increase embarrassingly with the time.

In comment on this generalization it is obvious in the first place that the operations are paper and pencil operations in the highest degree. Furthermore, the operations do not exist for answering the same sort of questions that can be asked about the energy of isolated systems. Suppose someone asks how we can know that we have *really* got the energy of the universe by this process, or how we know that the energy of the universe is really the sum of the energy of its parts and that there are not cosmic terms entering the energy of which we get no inkling in our measurements on isolated pieces of matter. There is no answer because there are no operations. That is, there are no *other* operations by which meaning can be given to "change of energy of the entire universe," and in the absence of other operations, our generalization, operational although it is, is nothing more than a convention. Whether such conventions play a sufficiently convenient part in the pencil and paper operations can best be answered by those people who are most concerned with the actual manipulations. If these people want this generalization and convention, there is no reason why the rest of us should not let them have it. In any event, the "energy of the entire universe" is a different sort of concept from the "energy of the matter in this room." It is different because it does not admit the same operations — either instrumental or verbal.

What we have just been doing and saying is part of the energy concept as used by every student of thermodynamics. I think it is obvious on reflection that what we have here is properly an extension of the energy concept as originally defined, and that there is no logical necessity that such an extension is possible. The fact that such an extension is possible is a matter for additional experiment; no physicist has, I believe, the slightest doubt that experiment does justify such an extension. We arrive, then, at the concept of energy as something of wider application than the operations by which we generated it. This being the case, it should be possible to devise some other way of generating it. There does indeed seem to be such a possibility, although the method is not very direct or simple enough to seriously use in practice. Given several different pieces of matter whose state parameters are completely determinable, and which may be independently varied through a wide range of conditions: These different pieces of matter are then allowed to react with each other in pairs under a wide variety of initial conditions and to various extents while enclosed in the impervious enclosure, and the corresponding initial and final parameters, after the pieces of matter have acquired homogeneity, are to be tabulated. Then it is possible by trial and error, if by no other method, to construct for each piece of matter such a function of its parameters of state that the sum of these functions for any pair of pieces of matter remains constant during any of the reactions which are the subject of study.

In fact, the sum of these functions remains constant when the bodies react in larger groups than pairs, or under a wider range of conditions than used in the construction of the energy functions. These functions are of course the "internal energy" of the various pieces of matter. Since the number of possible combinations in different conditions of the various substances is much greater than the number of the aggregate of possible energy functions, the energy functions are over-determined, and we are really "saying something" and have made a discovery in finding functions whose sum remains constant under the specified conditions. In fact, it is only this characteristic of "over-determination" that removes the energy function from the class of pure conventions — something defined only verbally by the first law itself. This remark applies also to the less general method of evaluating energy functions discussed on p. 69.

A remark of Poincaré is often quoted to the effect that if we ever found the conservation law for energy appearing to fail we would recover it by inventing a new form of energy. This it seems to me is a misleadingly partial characterization of the situation. If in any specific situation the law apparently failed we would doubtless first try to maintain the law by inventing a new form of energy, but when we had invented it we would demand that it be a function of an extended set of parameters of state, and that the law would continue to hold for all the infinite variety of combinations into which the new parameters might be made to enter. Whether conservation would continue to hold under such extended conditions could be determined only by experiment. The energy concept is very far from being merely a convention.

In evaluating the energy functions as above it is not necessary to introduce the notions of work and heat separately at all, except in the negative form that the boundary of the enclosure is to be impervious, which means merely that there is no flow of either heat or mechanical energy (work) across the boundary. The requirement that we know when there is no flow of heat or work is a much weaker requirement than that we be able to measure them when there is such flow. One way of being sure that the surface enclosing the system is impervious to the flow of heat or energy is to demand that the system be "isolated," and such isolated systems play an important part in many of the conventional thermodynamic discussions.

The business of "isolating" a system perhaps requires more discussion. What shall we say to one who claims that no system can ever be truly "isolated," for he may maintain that the velocity of electromagnetic disturbances within it or the value of the gravitational constant of the attractive force between its parts is determined by all the rest of the matter in the universe? To this all that can be answered is that this is not the sort of thing contemplated; even if an observer stationed on the surface enclosing the system is able to see other objects outside the surface and at a distance from it, the system may nevertheless still be pronounced isolated. It would probably be impossible under these conditions to give a comprehensive definition of what is meant in thermodynamics by "isolation," but the meaning would have to be conveyed by an exhaustive cataloguing of all the sorts of things which might conceivably occur to an observer stationed on the enclosing surface, with an explicit statement that this or that sort of thing does or does not mean isolation. In fact, whenever analysis is pushed to the limit, we are always driven from generalizations to an exhaustive cataloguing of all the things that have happened to us. But in practice anyone who has worked much at thermodynamics comes to decide so easily whether any particular system is isolated or not that he hardly feels the need for a definition. In general, the system would not be called isolated thermodynamically if anything happened to the observer at the surface which he ever under any circumstances associated with a "flow of energy."

The situation presented by the energy concept generalized in this way is awkward from the point of view of the physicist because of the difficulty of actually constructing the energy function for any specified piece of matter, but it is the sort of situation that the mathematician is familiar with and loves to talk about. A mathematician would doubtless say under these circumstances that an energy function "exists." He regards his task to be the manipulation of this function, but the manner of derivation of it he regards as none of his affair. In any given case he would start with the assumption that the energy function is "given."

The process suggested here for the generation of the extended energy function is very similar to the processes by which the concepts of heat and of calorimetry arose. For it was found when all possible pairs of bodies, at all initial temperatures, interact in impervious enclosures without the performance of mechanical work, that it was possible to assign coefficients (specific heats) to the bodies such that always

$$\int_{\tau_1}^{\tau} C_1 d\tau = \int_{\tau}^{\tau_2} C_2 d\tau.$$

The notion of "heat," and its conservation and therefore its thing-like nature comes at once.

Having now an energy concept of great generality, defined for circumstances in which the flow of heat and mechanical energy have no meaning, it is tempting to ask whether the original concepts of flow of heat and mechanical energy cannot be extended to have meaning in these more general circumstances? Imagine, for instance, a gas in which there is a temperature gradient rapidly expanding and pushing a piston before it. How shall we separate the heat from the mechanical work imparted to the piston, our answer being in terms of operations performed now at each physically infinitesimal area of the piston head? I confess that I now do not see what I would do to separate these two effects, but I do nevertheless have some idea of what I would like to try to do in the laboratory preparatory to finding an answer. For instance, I would like to know whether instrumental evidence of turbulence could be found on a scale much smaller than the scale of elements of area which give smooth values for pressure. Until such questions can be answered in the laboratory, the impulse to generalize the concepts of heat and work must be regarded as primarily a verbal impulse, but such impulses are nevertheless valuable because they suggest new instrumental programs. With regard to the probability that such a method of generalization will ever be found, it is to be remarked that if once a method of generalizing heat flow is found, the generalized flow of mechanical energy is thereby determined by the condition that together with the heat it shall give the already known generalized flow of energy. But originally heat and work had *independent* meanings and this should continue to be true in the generalized case. For this reason I am somewhat doubtful as to whether the desired separation into work and heat will ever be effected in the general case.

THE SEARCH FOR AN ABSOLUTE ENERGY

We have thus acquired an energy concept of great generality, but there are still restrictions that it will pay to emphasize. Notice in the first place that our operations give meaning only to changes of energy; we should talk about the energy of a body or system only when it is brought by some process from an initial to a final state, and should not speak of the energy of the initial or the final state by itself.* "Change-of-energy" should properly be regarded as a single verbal unit which it is meaningless to split into smaller verbal units, and which has application only to a "state-couple," not to a single state. If there is no "state-couple," no method of getting from the initial to the final state, the whole setting which gives energy its meaning evaporates and we have nothing left. If, for example, we may assume that it is a result of nuclear physics that there is no method of transmuting ordinary copper into iron, then it means nothing to ask what is the difference of energy between a gram of copper and a gram of iron. In any event it means nothing to ask what is the difference of energy between one and two grams of iron.

Let us analyze a little the reasons for the almost overwhelming verbal temptation to analyze "change-of-energy" into component parts. There is of course in the first place the verbal form itself, a combination of different words

* In partial extenuation of my early position with regard to localization of energy, referred to on page 25, it is to be noted that it was an "absolute" energy which I could not accept as being consistently localizable; this position of course is still maintained.

instead of a single word. But this is really no explanation at all, for what we are really trying to understand is why in the first place the concept was framed as a combination of words instead of as a single word. What is the reason for the verbal temptation to talk of "energy" as a thing in itself, in spite of the fact that we have given operational meaning only to "changes-of-energy"? The verbal temptation has its origin, I believe, in both the mathematical and the physical situations. On the mathematical side we have originally a function defined in terms of an initial and a final state and the path connecting them. It is a property of this function that the intermediate path drops out, leaving only the initial and the final states. It then follows at once for the functions of three states taken in the three possible pairs that $f(A,B) + f(B,C) = f(A,C)$. This is a functional equation, the solution of which is $f(A,B) = \phi(B) - \phi(A)$. That is, the function of the state-couple may be replaced by the difference of functions of the states separately. It is then natural to concentrate attention on the ϕ, which means verbally that we talk about the energy of a single state. But that this is not the whole story appears at once when we notice that the solution of the functional equation is not unique, and that a single ϕ by itself has no meaning, but acquires its meaning only when it is combined with another ϕ.

On the physical side we have, in the first place, a "change-of-energy" which is exactly localized. It is *inside* a definite boundary surface; the first law cannot be formulated without the use of such a surface. Its presence inside the surface can be detected and its amount measured by operations with instruments performed inside the surface, combined with certain paper and pencil operations. It is true that the paper and pencil operations may be complicated and require preliminary rehearsal, but that is beside the point. Parenthetically, the statement that the "change-of-energy" can be determined in terms of operations performed now inside the surface is simply the physical form of the mathematical statement that dE is a perfect differential in the state parameters. To see this, one has only to analyze what is involved in the notion of "state."

In the second place, the amount of "change-of-energy" localized inside the surface varies by its getting into the region across the surface. As it crosses the surface it can be detected and measured by sentries on the surface, and the increase inside exactly equals the amount which crosses. This statement is obviously true when the action at the surface can be represented in terms of a flux of heat and of mechanical energy as in the original formulation of the first law. When it cannot be so represented we maintain the picture by a paper and pencil extension employing electromagnetic radiation on a microscopic scale, or by the mathematical construction of a vector in terms of its divergence, as will be elaborated later.

In any event, we know of no phenomenon which is not consistent with the *conservation* picture for "change-of-energy." If one body gains energy, then another loses it, and in many cases we can trace the details of the transfer across the intermediate space. This conservation aspect is usually thought of as the most essential part of the energy concept, and the first law is sometimes formulated in the grandiose form "the total energy of the universe is constant," a statement which seems to me to be mostly a pure verbalism.

Having realized the inexactnesses and infelicities of the

ordinary usage, one is in somewhat of a quandary as to how best to proceed. One method frequently adopted is to invent a new technical nomenclature adapted to exhibiting the essential distinctions of meaning. This, however, may easily assume a forbidding aspect and in as well established a field as thermodynamics would, I think, have little prospect of adoption. I have, therefore, adopted the other more easy and more dangerous course. Instead of inventing a term for "change-of-energy" I shall continue to use the classical vocabulary of the physicist and speak of "energy," hoping that our discussion will have sufficiently awakened the critical consciousness of the reader to enable him to avoid the pitfalls. I shall later adopt the same course with regard to other terms also. To be successful this will require the active coöperation of the reader and also frequently his good will, for I cannot hope to be uniformly successful in using the old vocabulary without admitting the old implications. What is worse, I am afraid that I shall not always successfully avoid a certain degree of talking down to the reader, using a word in what I think will be the reader's understanding of it in the old inexact sense, whereas if I had the full courage of the conviction of the efficacy of my exposition I would be under no temptation. For such lapses I can only crave indulgence.

Granted now the attributes of localization and conservation, the similarity of "energy" to ordinary "matter" springs to the eye. This too is localized, and the amount inside a given surface can be determined by operations performed with instruments inside the surface. Furthermore, it is conserved, and the amount of matter inside a surface changes only by matter crossing the surface, and it can be detected as it crosses the surface by sentries posted on the surface with appropriate instruments. Certain aspects of the changes in which energy is involved can therefore be verbalized in the same way as changes involving the transfer of matter. Since simplification in verbalization is no less desirable than simplification in any other intellectual activity, such as mathematical calculation, the urge to verbalize matter and energy in the same way is quite understandable. But in their present forms there is at least one vital difference between the two. Not only does it have meaning to say: "There is now inside this enclosure more matter than there was an hour ago," but it also has meaning (instrumental and operational) to say: "There is now inside this enclosure just so much matter." The latter statement obviously cannot be made about the "change-of-energy" defined in terms of the operations of the first law, but it would be most convenient if it could be. Furthermore, it hardly seems right that, in a given frame of reference, we can verbalize about kinetic energy and electromagnetic energy in a different way than we can about the generalized "change-of-energy" of thermodynamics which has the same dimensions. For we can say of the former as we can of matter: "There is now inside this enclosure just so much kinetic energy of mass motion, or just so much electromagnetic energy stored in the field."

A very natural and perhaps profitable reaction to a realization of the situation is to ignore the differences and to embark on a program of physical exploration to find whether there may not be new sorts of physical phenomena which would permit completely analogous physical operations for matter and energy in general. The psychological attitude of the physicist in embarking on such a program will be to a large extent a matter of individual temperament — all gradations are possible between extreme mysticism and the most cynical hard-boiledness. Some physicists have almost a religious conviction that it *must* be possible to carry through such a program; such convictions may have their economic justification in scientific society just as much as similar convictions in ordinary society.

The urge to treat energy as a thing has deep historical roots; it goes back at least to the early experiments on calorimetry by Black and the invention of the caloric fluid. The utility in the historical development of calorimetry of the concept of a caloric substance that is conserved hardly needs to be emphasized. But of course the program of experiment based on the notion that there must be a definite quantity of this fluid in every substance failed. It was the failure of this program that fortified the concept of thermal energy as random kinetic energy of motion of the concealed small parts of the body whose existence we find so verbally convenient. This concept prepared the way for the notion of the interconvertibility of kinetic and thus of all mechanical energy into thermal energy, and at once offered a perspicuous explanation of the conservation property. But at the same time the prospect had to be given up of finding a complete analogy between energy and a "thing" because of the way in which the ordinary kinetic energy of mass motion changes when the velocity of the frame of reference changes. The mass of a stone is unchanged (except for relativity effects) when it is measured in a frame of reference moving with respect to the original frame, but its kinetic energy is drastically altered. The obvious impossibility of forcing ordinary kinetic energy into the required mold has I think produced the logical effect upon those who have been mainly concerned with mechanics, and there is little or no temptation to think of energy as a thing among mathematicians concerned with classical mechanics. Historically, the immediate effect on the physicist also of his discovery that heat is a "mode of motion" appears to have been to free him from the temptation to think of energy as a thing. But the physicist is concerned with a much wider range of phenomena than the pure mechanics of the mathematician, so that the temptation to treat energy as a thing returned later and was not resisted; in fact the presence of temptation was not even recognized. The ordinary thermal properties of a substance are, except for relativity effects which are very small, independent of the motion of the frame of reference. The temperature of a body is not altered by setting it in motion, and the amount of thermal energy required to bring it from one temperature to another is similarly independent of the velocity of the frame of reference in which it is measured. This at first sight might seem paradoxical in view of the dependence of kinetic energy on the velocity of the frame, but the explanation is that thermal energy is *random* kinetic energy, and this part of the total kinetic energy is unaffected by motion of the frame of reference. We have here, then, for the change-of-thermal-energy the same sort of independence of the velocity of the reference frame that we have for mass, and the verbal impulse is strengthened to assimilate the two.

As far as purely thermal phenomena are concerned the urge to verbalize about energy as a "thing" cannot well have any other issue in the form of a laboratory program than the attempt to find some natural and unique zero point from which to figure changes of energy. Now there is apparently such a natural zero, namely the state of the body at the absolute zero of temperature and free from the action of external forces, or closely enough for most practical purposes, simply at atmospheric pressure. Such a zero point has often been proposed and used. The advantages, however, are disappointing. In the first place, there is no simplification in any of the instrumental operations; we have to continue to take readings over a series of temperatures, which of course gives the opportunity to take the difference between any two, as is demanded in the definitions. In the second place, many substances exist in two or more forms at zero absolute, with energy differences between the forms. In the known cases we may specify that the origin shall be taken from the form with the smallest energy, but there can never be an answer to the objection that the substance may have unknown forms of lower energy, for there is no process known at present by which the possibility of new forms may be ruled out. With this possibility always staring one in the face, it is difficult to retain our conviction of the absolute significance of our present zero point, or our satisfaction with it.

If we had been restricted to purely thermal phenomena I think the verbal impulse to treat energy like a thing whose absolute amount has physical significance could not have survived the various incongruities. The strength of the impulse comes from the electromagnetic side. The instruments for determining the energy localized at a point in the field are instruments for measuring E and H; the paper and pencil operations combine the instrumental data into the formula $(E^2 + H^2)/8\pi$. Although the detailed data of the instruments are different when the frame is set in motion, nevertheless the changes in H compensate those in E in such a way that the sum of the squares is, except for small relativity terms, unaffected by the motion of the frame. Furthermore, the instrumental operations have the same sort of natural zero from which to start as the operations for determining total mass. The natural instrumental zero for electromagnetic energy is the condition in which E and H are both zero, and calculated from this zero it is just as simple and natural to talk about the "absolute" energy as it is to talk about an absolute mass (independent of the velocity of the frame). From the point of view of the paper and pencil operations there is not the same natural uniqueness, because to the $(E^2 + H^2)/8\pi$ determined by the instruments there may be added any arbitrary constant with no change in any consequence testable in the laboratory.

The idea that mass might be of electromagnetic origin was an early and an attractive one and received much encouragement from the mathematics of the energy stored in the electromagnetic field. A study of the transformation properties of the energy led to the notion that "mass" might vary slightly with the motion of the frame of reference and so led to the transformations of relativity. Later, in the hands of Einstein, the similar behavior of all mass was postulated, and the thesis that all mass had to be of electromagnetic origin was given up. But the electromagnetic relation between mass and energy is retained, and we now have the universally accepted doctrine of the equivalence and interconvertibility of mass and energy. Here at last we have the thing that we have been searching for so long: a unique value in any reference frame for the energy of a system and a simple physical significance for the thermodynamically undetermined "constant of integration" of the energy equations. The total energy of a system is then simply proportional to its total mass, and the variations of energy which are the subject matter of thermodynamics are mirrored in changes of the mass and could be so determined if we could make our measurements accurately enough. It is true that as far as thermodynamic phenomena are concerned an accuracy would be demanded fantastically beyond present instrumental possibilities. But the paper and pencil scheme is complete, and wherever the experimental verification can be attempted, as in nuclear physics and all transmutation experiments, the check has been spectacular.

There are two observations which I would like to make with regard to this situation. In the first place, the idea of the equivalence of mass and energy is not a unique consequence of the relativity point of view, although the proof of the equivalence is often cited as the brightest jewel in relativity's crown. The equivalence is a consequence of the laws of ordinary mechanics and such elementary properties of the electromagnetic field as the pressure of radiation. A simple proof can be given by studying the changes of momentum of a box in which radiation is bouncing back and forth between mirrors, and the result was known before relativity theory was formulated. This, I think, is important to realize, because the experimental verification, which is now being given so spectacularly by transmutation experiments, is often held to check the theory of relativity. Relativity theory includes other and to my mind much more questionable assumptions than are necessary to deduce the equivalence of mass and energy, and a verification of the one is by no means a verification of the others.

The second observation is that the fundamental conditions under which the energy concept arose and which gave it meaning still hold. We are still (and must be because of the operations) concerned only with processes from an initial to a final state, and the energy concept has no meaning apart from a corresponding process. One cannot speak of the equivalence of the energy of mass and radiation unless there is some process (not necessarily reversible) by which one can get from mass to radiation. It means nothing to say that the mass of the electron indicates that it has stored up in it a certain amount of energy unless there is some process by which the electron can be annihilated with the appearance of energy in some form in its place. The same sort of statement applies to nuclear transformations. I think this is recognized with complete clearness by any physicist who is actually working in the laboratory in this field. But I cannot help feeling that in some places there is a trace of mysticism left in the verbal impulse to say "matter and energy are *really* the same thing."

The "operational dimensions" of mass and energy are different, and to that extent they cannot be "really" the same thing. The word energy is properly used only when we are concerned with making a transition from one configuration to another, and the physical interest of talking about energy is that it is concerned with the transition process. It is true that the aspects of the transition process which interest us may be uniquely connected with measurements made on the initial and final configurations only, and we may if we like say as a shorthand manner of expression that these initial and final measurements are measurements of "energy." But we must not forget that this is a shorthand expression only, and that, unlike mass, no single measurement of "energy" has significance by itself, but that it acquires significance only when it is combined with a second similar measurement.

FLUX OF ENERGY IN GENERAL

So much for the possibility of treating the energy of a substance like a "thing." What now about the "flux of energy"? can this be treated like the flow of a thing? Notice in the first place that there is a uniqueness about the instrumental operations — that is, a natural and simple zero — such as we failed to find in seeking for a natural zero of thermal energy. This uniqueness of instrumental operations connected with the flux pertains to thermal, electromagnetic, and mechanical flux, all three, whereas with regard to energy density we found such a natural zero for electromagnetic and for gross mechanical, but not for thermal energy. The instrumental operation for getting thermal flux involves getting the thermal gradient, and the natural zero is when the gradient is zero. The electromagnetic flux is determined in terms of the Poynting vector, and this means getting E and H and taking their vector product, with a natural zero when E or H is zero. The flux of mechanical energy demands determining the mechanical stress and the bulk velocity, both of which have natural zeroes. Perhaps if the fundamental difference with regard to an instrumental zero between flux and density of energy had been noted earlier in the historical development we would not have had the verbal impulse to analyze a common energy out of "flux-of-energy" and "change-of-energy."

Thermal and electromagnetic flux are, except for small

relativity terms, unaffected by motion of the frame of reference (except the part of the electromagnetic flux arising from motion through a static energy field, which is like mechanical energy flux). The flux of mechanical energy is, on the other hand, drastically altered by motion of the frame. In this respect the flux is like the mechanical energy itself, and is as to be expected. The uniqueness of the instrumental operations is therefore perhaps on the whole favorable to the treatment of flux of energy like flux of a thing. But as in the case of energy density, the paper and pencil operations destroy this uniqueness, because to any instrumentally determined flux may be added any divergenceless vector, with no change in any of the laboratory consequences. It is a paradox that we can find a physical and instrumental uniqueness for the operations of our sentries at points on the bounding surface, but when we have it, we can make no use of this uniqueness in formulating the first law. We wonder whether the uniqueness can have the physical significance of which we were convinced when we started on our program, and we wonder how much purely verbal bedevilment there may be here.

Perhaps the crowning difficulty in trying to carry through the "thing" point of view comes when we try to analyze flux into a product of a density and a velocity. This can always be done for the flux of substance. But, as we have already seen, it is impossible for the flux of mechanical energy, and therefore impossible in general, because we cannot find any velocity for whatever it is that has the density which transforms in the right way with the motion of the frame of reference.

The whole situation with regard to flux changes when the mutual interaction of the elements of the system is so complex that we cannot give meaning to flux of thermal and mechanical energy. As we have seen, we have a general method for assigning an energy function to the elements of any body by the method of unlimited trial and error in all possible combinations with all possible other bodies. Given now any complex system of bodies, whose energy function is known as a function of its parameters, undergoing reaction with each other under conditions so complicated that it is impossible to assign meaning to flux of heat or work between the elements of the system, since we still have conservation, we would like to verbalize in the same way as we did before. I think that our fundamental verbal demand here is that we should be able to say: "The additional energy now inside this region above that which was there a moment ago is equal to that which has entered the region across the boundary plus that which has been created inside." This verbal demand I think we will always make, not only of "energy" but also of any other sort of "thing" which we can localize within the surface. Personally, I do not see how I could verbalize the situation in any other way. This is a purely verbal matter, irrespective of the operations by which I obtain the flux and the accumulation of energy. I would still make the same statement whether the energy flux is given only by instrumental operations made now at the point of flux or whether the instrumental operations had to be performed at other points of space and over an interval of time, that is, whether or not they have "physical reality." I feel myself under a verbal compulsion, from which there is no escape, and I believe from observation of my fellows that they feel similar verbal compulsions. It is important to recognize that such verbal compulsions are at work, molding our formulations both in mathematics and in our less rigorous linguistic usages.

The formulations we make under such compulsions cannot have the significance of their naïve face value. Such a verbalization is empty until some instrumental content is given to the terms which are connected by the verbalization; in this case there are three such terms: "additional energy," "energy which has entered," and "energy which has been created." Since we have the generalized procedure for assigning an energy function, the term "additional energy" does have the required instrumental content. But we are not so fortunately situated with regard to the other two. I know of no attempt to formulate what would be the earmarks of the creation of energy inside a closed region, and until such a formulation is successfully made our verbalization is nothing more than a conventional definition of "creation," even assuming that we can give an instrumental meaning to "flux-of-energy." Of course we seldom feel impelled to consider this general situation because of our conviction that we can always find an energy that is conserved, that is to say, never "created." If we assume conservation, our verbalization becomes "The additional energy now inside this region is equal to that which has entered it across the boundary." Still we are "saying" nothing — that is, we are only defining what we mean by "enter across the boundary" — unless there is some independent instrumental meaning that we can give to the flux. The purely conventional character of a flux which is fixed only by the verbalization is at once obvious from the mathematical analysis. The flux must be such a vector that its negative divergence is everywhere equal to the rate at which energy is there accumulating, something which does have independent meaning. The mathematical problem is simply to determine a vector, given its divergence everywhere. The problem has a solution, and furthermore a solution which is unique except for limitations which do not concern us. But in the absence of physical operations, such a solution has no further significance. In fact, if I choose to say that half the energy appearing in each element of volume was created within it, and only half entered across the boundary, I would still get a unique solution for the flux. The mathematics gives a unique flux no matter what the divergence, and whether or not the integral of the divergence over the universe is zero, that is, whether or not there is conservation in the large. With such a conventionalized flux I cannot even give any meaning to the statement "energy is conserved in *detail*, in the interaction of all the elements of any system." I can only say that energy is conserved on the whole in the interaction of systems in impervious enclosures, because it was only by means of the interaction in impervious enclosures that I was able to assign an energy function in the first place.

If it were not for the existence of impervious enclosures we could not give any meaning to conservation. The operational meaning of "impervious" is an interesting subject for analysis. I believe the situation is that if the enclosure is surrounded by the proper material we are satisfied to say that it is impervious. That is, our operations have changed from continuous local operations by the sentries to operations made once and for all to assure ourselves that the material of the boundary has certain properties. The operations which satisfy us that a material will not transmit generalized energy are, I think, merely that it will never transmit either heat or mechanical energy in those simple cases where $\vec{q}$ and $\vec{w}$ have meaning. The situation is, therefore, far from being logically satisfying.

From the macroscopic, phenomenological point of view we can, therefore, always save the situation by talking about a generalized flux of energy when the situation is too complicated to analyze into thermal and mechanical flux. But this will remain a pure convention until the instrumental operations are discovered which will give meaning to such a generalized flux, and of the possibility of this I think there is no present indication. However, from the atomic, microscopic, paper and pencil point of view, which knows no temperature or thermal phenomena and which views everything as "mechanical" (electromagnetic on the small scale), it does make sense to talk about the flux of energy in general and about conservation in detail, because there are independent operations for the flux (the Poynting vector), and when the flux and energy increments are expressed in terms of these independent operations and substituted in the general verbal form, the remaining term for created energy is identically zero. As always, the justification for this paper and pencil procedure is that when by means of it we emerge again into the macroscopic domain we find ourselves at the right place. The concept of generalized energy flux thus has its utility in spite of its macroscopic conventional character.

We are happy that the flux, which might play a purely formal role, proves to have a "physical" significance and can be specified, at least in a great many situations, in terms of operations with instruments. Our satisfaction is further increased when we observe that the flux plays still another role which has not yet been mentioned. The flux of energy across the boundary of a region gives the rate at which energy is increasing within the region, so that, given the flux everywhere, we can predict the future of the energy localized at every point of the system, so long as the flux remains constant. In an ordinary mechanical system velocity plays a similar role in enabling us to predict the future position of the parts of the system. In either case the ability to make such predictions is demanded by all experience; if the future is determined by the present, then we should be able to make measurements in the present that will enable us to predict the future. In a mechanical system these measurements are of the velocities, and in a thermal system they are of the various kinds of flux.

Having accepted this point of view, we might on first reflection find it strange that there is no close correspondence between thermodynamics and ordinary dynamics; there is nothing in thermodynamics corresponding to the equations of motion of mechanics. The explanation is merely that the situation is so simple with regard to thermal effects that there is no need for an elaborately developed discipline; this simplicity results from the fact that there is no such thing as thermal inertia corresponding to the inertia of mechanical masses.

In any event, our physical intuition, whatever it is worth, would be outraged if we could not perform instrumental operations in a thermal system which would enable us to predict its future; this the flux does. It is not logically or intellectually necessary that the instrumentation which puts us in a position to predict be of the character of the instrumentation by which we measure flux. However, certain other physical intuitions find it very congenial that it is of this character. There is evidently a close connection with the physical intuition typified by the classical English physicists to whom action at a distance was repugnant and who demanded propagation of effects from one proximate part to the next. The "reality" of such a propagation through intermediate spaces is witnessed by the possibility of giving an instrumental specification of the flux. It is pleasant that things fit in so nicely with the demands of this point of view; I think, however, that if the facts had been otherwise we would not have despaired of being able to make shift to deal with the situation with logical consistency.

In all this consideration of flux, the verbal character of what we do is particularly prominent. By what necessity do we verbally analyze flux into pieces, and pick out a part to which we give the name of energy, saying flux OF *energy*? The general nature of the situations to which the first law applies is not different from that of any situation in physics which we bring under a general causal point of view; we merely have a *correlation* between something measureable which happens inside a region and something measureable that happens elsewhere, which may be in particular something that happens at the boundary. We ought to be able to be content with a mere statement of this correlation without having to go further and say that the same *thing* can be found in action at the boundary that can be found inside. It seems to be possible by a slight verbal rearrangement to minimize the unfortunate implications of our customary verbal analysis. Instead of saying "flux of thermal energy," "flux of mechanical energy," and in general "flux of energy," we might simply say: "thermal flux," "mechanical flux," or "energetic flux."

This is probably as good a place as any to comment on one other feature of our verbalizing which has obviously played an important role in shaping physical theories. The constant impulse to introduce things, as shown by our necessity for thinking of energy as a thing or our inability to speak of a flux without saying flux of something, is certainly fortified and made natural by the grammatical structure of our language which turns verbs into nouns. Consider, for example, a flywheel with axle around which a cord is wrapped attached to a weight. The flywheel is slowing down and the weight is being wound up by the cord. One aspect of this system may be described about as directly and noncommittally as can be imagined by saying "the flywheel works on the weight." But the English language is such that this is grammatically equivalent to saying "the flywheel does work on the weight," which can be further transformed into "the flywheel gives work to the weight" or "the weight receives work from the flywheel." Because these forms are grammatically equivalent it is easy to think that they are also physically equivalent, with a train of consequences some of which we have been examining. Most of the European languages appear to have a similar grammatical structure; it is interesting to speculate what might have been the characteristics of a physics which had grown up in an entirely different linguistic medium.

FORMULATION OF THE FIRST LAW WHEN THERE IS MATERIAL OR ELECTRICAL FLUX

We ought now to be able to let fall some of the remaining restrictions in our formulation of the first law. We required at first that the sentries at the boundary of the region should observe nothing which they could not describe as flux of heat or of mechanical energy, and later we supposed that we could put into the mouth of the sentries a report of what we called "flux of energy," although this is at present mostly a convention when we get outside the paper and pencil domain. What we are now concerned with is what to do when the sentries report other kinds of occurrence, such as the flux through the surface of a stream of matter or of a current of electricity. We still ought to be able to deal with the situation in much the same way as before. We are able to say "The matter now inside the surface is that which was originally there plus that which the sentries have observed to enter across the bounding surface" and, if it is conserved, we ought to be able to make the same sort of statement with regard to energy. In fact, as we have already seen, this is what "conservation" means. What operations shall we perform on the report of the sentries in order that we may say this? It is in the first place obvious that all we can require of the sentries is a complete report of the results of every operation in their command, which they can perform at the boundary "now." At least we would be very unwilling to give up this requirement, which is in accord with our instinctive demand for the "physical reality" of the flux across the boundary. I do not know to what extent this instinctive demand is properly to be described as "verbal"; I think it is one which we would find it very difficult not to make and which reaches pretty far down into the pattern of our thinking. It is at any rate a demand which it turns out we are able to exact in the situation before us.

If now there is a flux of matter across the boundary, we have merely to say that the matter convects energy with it across the boundary. We shall get the right result if we take as the zero from which to measure the amount of convected energy the same standard state of the body as that which we use for measuring the changes of energy of matter which may remain completely inside the surface. The sentries are required to observe the amount and kind of matter which crosses the boundary and its state as it crosses. None of this required information involves anything which the sentries cannot do with conventional physical instruments while stationed at their posts. The precise way in which the clearing house shall use the information of the sentries demands, as always, preliminary rehearsal.

If the sentries observe the flow of electrical current across the boundary then again it is possible to say: "The electric current convects energy with it into the region." Just what observations the sentries must make in order to specify this convected energy can be determined only by detailed examination, but it turns out that it can be done in terms only of instrumental operations by the sentries at their posts. Among other things it involves a measurement of the amount of current, as is to be expected, and in addition certain of the properties of the conducting material in which the current is flowing, such as its Thomson coefficient, and also of the temperature gradient. In this way a complete, somewhat novel, and suggestive description of what happens in a thermoelectric circuit can be obtained. It is possible to carry through a consistent verbalization of the entire situation by means of the concept of two different kinds of electromotive force, a "driving" and a "working" electromotive force. The details will be found discussed in my book on the *Thermodynamics of Electrical Phenomena in Metals*. The point of view I believe to be fruitful, for it seems to me that the one kind of electromotive force recognized in the conventional electrical analysis of macroscopic phenomena in conductors is not adequate to permit a complete or consistent verbalization of the thermoelectric situation.

There is one feature of the way in which physicists have received the suggestion that two kinds of electromotive force may be necessary in describing thermoelectric phenomena which is illuminating enough to comment on. My contention that conventional methods of describing electric phenomena in metals are inadequate was the outcome of an entirely phenomenological analysis, all the instrumental operations being the ordinary macroscopic operations of the laboratory, and any paper and pencil operations being no different in character from the paper and pencil (verbal) operations of classical theory. The result of my analysis was naturally suspect, because it impugned the adequacy of the classical scheme of description which had hitherto been accepted without question, and it was to be expected that no physicist would accept my conclusions without his own independent scrutiny. It is the character of this scrutiny which appears to me significant. Instead of examining the macroscopic argument to see whether it was cogent or not, it was, I suppose, the first impulse of everyone who concerned himself with it to say: "This cannot be right, because there is no place for it in the electronic picture." Then later those few who bothered with it further did discover the somewhat obscure place for it in the electronic scheme, and then reversed their first judgment and said "It is probably all right because I can see what it means in terms of electrons." The complete confidence of most physicists in their atomic analysis is I think quite typical; we have already seen that it colors the current attitude toward thermodynamics and kinetic theory, but it overlooks the essential fact that logically the microscopic picture had its origin in the macroscopic. Any suspicion of the adequacy of our macroscopic analysis should at once arouse suspicion that our microscopic analysis is also inadequate.

GRAVITATIONAL ENERGY

The situation with regard to gravitation — gravitational forces and gravitational energy — has features which are worth special examination. In the first place, there is for

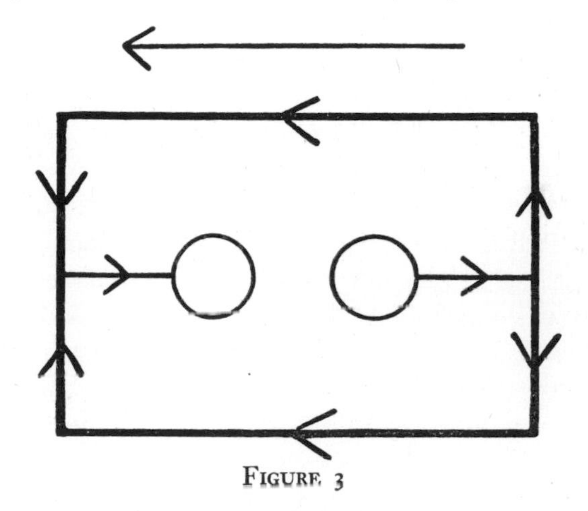

FIGURE 3

gravitation nothing corresponding to the Poynting vector for the flux of electromagnetic energy. This corresponds to the physical fact that there is no known special effect in the neighborhood of moving gravitating masses analogous to the magnetic field which surrounds moving electrostatic charges. Nevertheless, the need for something analogous to the Poynting vector is distinctly felt in certain situations. Consider, for example, the system indicated in Figure 3. Two equal gravitating masses, in empty space, are prevented from approaching under their mutual gravitational force by two cords attached to a frame as indicated. The whole system, masses and frame, are in uniform motion with respect to the frame of reference. The cords and the members of the frame are the seat of ordinary mechanical stresses, and since they are in motion, they are the path of a flux of ordinary mechanical energy. This flux of mechanical energy is as indicated in the figure, out of the rear mass, through its cord into the frame, along the members of the frame and into and through the forward cord into the forward mass. The net effect is a continuous transfer of energy from the rear to the forward mass. But the system is in uniform motion, and by all that's holy such a system must continue indefinitely in such a state of motion with no internal change, in spite of the continual accumulation of energy in one mass and the impoverishment of the other mass. This, by the general principle of the equivalence of mass and energy, should at least result in a continual change in the relative magnitude of the masses and a shift of the center of gravity. The only obvious way of saving the situation is to postulate a compensating flux of energy in the gravitational field in "empty" space from the forward mass into the rear mass.

The need for such a gravitational Poynting vector seems to me so imperative that I should think one might make the postulation of its existence the basis of a laboratory program of search for new effects in the neighborhood of moving gravitating masses. One may be sure that such effects are terribly small; Oliver Lodge and perhaps others have made unsuccessful search for such effects in certain special situations. It is further to be said that general relativity theory does have a term in its equations which to some extent plays the role of a Poynting vector. But this vector has no simple relation to the physical set-up, and even in the simple case of two gravitating masses just considered is so complicated that it is practically impossible to calculate it. "Physical intuition" makes me feel that there should be some way of treating gravitational problems in which there is a simple Poynting vector which plays a fundamental role.

Granted now that there is no flux vector for gravitational energy, let us see how we are driven to localize the energy and what account we are forced to give of it in systems with gravitational forces. Our fundamental definitions and operations by which energy received meaning still demand that the energy be localized, and if we possibly can we are going to stick to our requirement that the operations by which localization is effected be such that we can ascribe "physical reality" to the localized energy. Consider the simplest possible case of a body near the surface of the earth freely falling from initial rest. Apply the first law to a closed surface so drawn as to include the initial and a later position of the body. By hypothesis there is no flux across the boundary, so that the total energy localized inside the surface must remain constant. This total energy must be given in terms of the instrumental operations that can be performed on the contents of the surface initially and at the later instant of time. There are a great variety of instrumental operations that could be performed on the region inside the surface, but after a considerable experience with different sorts of falling body we would come to the conclusion that there are only two sorts of operation that are pertinent to the situation, an operation which measures the vertical velocity of the body, and an operation which measures the height of the body above the floor of the region in which it is. Now we already know all about kinetic energy of mass motion from our elementary mechanics, so that if we are to have a consistent whole we are forced to say that the kinetic-energy part of the whole energy inside the box has increased. It must be, therefore, if our scheme can be carried through, that there is some compensating change in the energy regarded as a function of the other parameter, the height above the floor of the box. This of course is the case; the compensating energy is the "potential" energy of position given in terms of a simple instrumental measurement of position, and the sum of kinetic and potential energies is constant.

The localization of the kinetic energy is obviously in the moving mass. One is also driven to similarly locate the potential energy in the falling mass. The argument comes by considering the situation with enclosing surfaces of different shapes, first a loosely fitting surface and then a cylindrical one fitting tightly as a sheath over the actual path of fall. Since there is the same increase of kinetic energy in all these cases, and also the same sum, one cannot localize the gravitational energy in the space outside the masses, but is driven in the limit to localize it entirely in the falling mass itself. The same argument shows that the other instrumental results which can be obtained in the space surrounding the falling mass are not pertinent, in particular the gravitational field of the earth and the much weaker gravitational field of the falling mass, which at a fixed point of the box changes with time as the mass falls past it, and to which one might at first be inclined to attach a special significance.

We have thus been able to maintain our thesis with

regard to the general formulation of the first law, for we have been able to set up rules for connecting the results of measurements made now with instruments inside the closed surface with the results of other instrumental measurements made at all points of the surface (these latter give a result identically zero) in such a way that conservation is preserved. I think that most physicists do not realize that this is possible in the simple non-relativistic gravitational case (for the remark is often made in connection with general relativity theory that gravitational energy is not localized), and not realizing the possibility are not sufficiently disturbed at the inferential impossibility. However, accepting now the possibility, it is questionable how much physical satisfaction a physicist with the sort of physical intuition of a Kelvin or a Maxwell can take in it. For one thing, such a physicist would want to ask by what mechanism potential gravitational energy gets transformed *in situ* into kinetic energy, and there seems to be no answer. Another difficulty will be elaborated in the next paragraph.

The gravitational case and the electrostatic case are entirely different with regard to the localization of energy; in the former we have to "say" that the energy is all potential energy of position, localized in the mass itself, and in the latter we say that the energy is localized throughout the field in amount $E^2/8\pi$. But it is well known that in the case of the electrostatic field there is another method of localizing the energy which is mathematically identical, namely the energy $\frac{1}{2}V\rho$, where V is the electrostatic potential, ρ the charge density, and the localization is in the charge. Why is it that to the latter localization in the electrostatic case, which is the analogue of the localization which we have been forced to give in the gravitational case, is ascribed only a mathematical significance? The reason, of course, is that in electrical systems we do have another phenomenon, the magnetic field. This can be detected by special instruments, not at first electrostatic in character, and always is present when electric charges are set in motion, that is, when they are moved from place to place. For ordinary electrical motions this magnetic field is so small that it is extraordinarily difficult even to detect, as shown by the special mission to Paris on which Rowland had to send Pender to convince the doubting French physicists. But it is not the magnitude which is important, it is the mere existence. If the motion is made twice as slowly the magnetic field is one half as great and the Poynting vector one half as great, but the time is twice as long so that the total transfer of energy from one configuration to the other is independent of the velocity of the displacement, and therefore the details of the initial and final localization of energy are unaffected by the velocity of the displacement. We are still driven to localize the electromagnetic energy in the field, then, even when the magnetic field is so small as to be hopelessly below the possibility of observation. But if we had never observed any magnetic field at all, if Rowland had not been skillful, we would still be localizing electrical energy in the charges and we would have been entirely satisfied with all the consequences that we could deduce from such a localization.

The situation thus disclosed does not "feel" right; the localization of a thing with the "physical reality" which we have been insisting on ascribing to energy should not be drastically altered or be alterable by the discovery of the existence of an effect so minute that it may have evaded the search of generations of scientists. It seems to me that the conclusion must be drawn that the notion of "physical reality" — what we imply when we "say" that something has "physical reality" — involves still other demands, perhaps some of them nebulous, beside those which we have hitherto exacted, and that it is by no means possible to maintain that energy has "physical reality" in the fullest sense.

SUMMARY — THE GENERAL NATURE OF ENERGY

In concluding this chapter I summarize briefly the sort of thing that the energy is which is the subject of the first law. It must be localized because of the method of formulation of the first law. At first it is defined only for processes for which flux of thermal and mechanical energy have meaning. Under these conditions there is a conservation law. Later the energy concept may be generalized, but only by the use of the conservation law and the notion of boundaries impervious to the passage of this generalized energy, a procedure which suffers from the logical reproach of a mildly vicious circularity, but which leads to no practical inconsistencies.

The energy of the first law is properly not a "point" function, although often so described, but is a function of a "state-couple" since it is defined in terms only of an initial and final configuration of a system. In this it is unlike other things with the same "dimensions" which are also referred to as "energy," such as the kinetic energy of mechanics, which is a point function in that it has a unique value in terms of the instantaneous velocities of the system. We are interested in energy only when we want to go from one configuration to another; as long as we remain fixed it is of no concern to us.

There has been a very important verbal element in our formulation of the first law. There is no single instrument for measuring changes of energy; but it is nevertheless possible, given the material, and after suitable rehearsal, to give the instrumental operations by which the energy of that particular material may be assigned, and so to assign the energy as a function of the state parameters. The different instruments and the different procedures which we use to catch all the different "forms" of energy have nothing very obvious in common as initially presented to us. What they have in common is a common purpose and a common use in our paper and pencil manipulations. We lump all measurements together and call them measurements of energy change if we add them to obtain the total dE of our formulation. The common name helps to keep in mind the common use in the face of the instrumental diversity; we have found it convenient to invent such a name because the use is so frequent. Verbal transformation has similarly been active with what the sentries do at the boundary. Because there is a relation of equality (that is, equality in the numerical results of our paper and pencil manipulations) between the measurements of the interior instruments and the measurements of the sentries on the surface, we call them by the same name, and speak of heat and work as forms of energy, although such an "equivalence" might have been the last thing we would have suspected from a naïve inspection of what was done with the instruments.

In the simple situations which originally give the energy concept its meaning it is possible to talk about a flow of heat and of mechanical energy. These flows are given uniquely by instrumental operations, but this uniqueness is not utilized in the resultant thermodynamic energy, which depends only on the divergence of the flow. In the more general case, when it is impossible to determine instrumentally a flow of heat and mechanical energy, we can still talk about a flow of generalized energy, but from the macroscopic point of view this is a convention without complete instrumental significance. There is, however, utility in the concept from the paper and pencil point of view of atomic operations. Going further, it is not possible in general to analyze a flow of energy into two factors, one of which has all the properties of the density of a material thing and the other of which has all the properties of the velocity of a material thing.

It thus appears that when we consider the entire picture energy is pretty hybrid and nondescript if we want to set up a parallelism with ordinary material things. Hence we should not attempt to set it up. But I think that many physicists do undeniably feel the temptation to set up this parallelism. The temptation is I think largely verbal in origin. I would suggest that the temptation may be largely avoided by a verbal device. If instead of talking about "energy" one always says "energy function" I think many of the implications of the conventional point of view will be avoided. For a function is obviously a paper and pencil affair without material implications. And the paper and pencil element in the energy concept is inextricably interwoven with what we should like, if we could only make it mean anything, to call "purely physical." By talking of an energy function we would keep before us the man-made nature of the whole thing, and this is the essence of what we have been saying.

CHAPTER II

THE SECOND LAW OF THERMODYNAMICS

THERE have been nearly as many formulations of the second law as there have been discussions of it. Although many of these formulations are doubtless roughly equivalent, and the proof that they are equivalent has been considered to be one of the tasks of a thermodynamic analysis, I question whether any really rigorous examination has been attempted from the postulational point of view and I question whether such an examination would be of great physical interest. It does seem obvious, however, that not all these formulations can be exactly equivalent, but it is possible to distinguish stronger and weaker forms.

ISOLATED SYSTEMS SEEK A DEAD LEVEL

All the various formulations of the law have as background a certain universally observed characteristic behavior of all physical systems. It is common knowledge that a great many systems, when left to themselves, do not remain inertly in the condition in which they are first observed, but a sequence of changes occurs. The violence of these changes may show all gradations from what occurs in a keg of powder into which a lighted match is falling to the running down of a watch. All these sequences have this in common, that after initial stages, which may be of various degrees of violence, the system settles down more or less asymptotically into a dead level of quiet and uniformity from which it can be aroused only by some sort of action from outside — that is, only by breaking the condition that the system be isolated. This characterization of the sort of thing that happens in isolated systems is an exhaustive characterization in the sense that no actual system is not of this kind. Included here is the special case that the system has already reached its dead level; such systems are already inert, and nothing is observed to happen in them. The only other sort of systems which could very well conceivably occur would consist of those in which changes continue to occur spontaneously with no tendency to die out. A special case of systems of this sort would be those in which a round of changes occurs cyclically over and over again. Such a round of changes occurs ideally in any purely mechanical system, as in a flywheel rotating forever on frictionless bearings. Certain very general considerations suggest that it is the sort of thing that might be expected at first glance of all systems. For most systems are causal systems in the sense that when their parameters of state are completely given their future history (while the systems are isolated) is uniquely determined. Although this way of putting it is hopelessly verbalistic, as can be seen on asking what is meant by *completely* giving the parameters of state, the verbalism is innocent enough, and means only that as a matter of fact it is possible to make a sufficient number of statements about any actual system so that when in the future we can again make those same statements the further history is repeated. This is never rigorously the case, but is an ideal limit to which we have often approached very closely, and which we shall assume in our paper and pencil operations we might approach indefinitely closely if we took the trouble. We assume, then, that all systems are causal systems. Further, we assume that they are finite systems, which means merely that their parameters of state are capable of taking on only a finite range of values. Then whenever in the sequence of changes the original values of the parameters are recovered, the entire sequence of changes from that point on must retrace its original course. This is the sort of thing that might be expected, but it does not occur. Its occurrence is prevented by something in all systems which is the analogue of friction in the rotating flywheel. The kinetic energy of the flywheel is converted into thermal energy by friction. This first shows itself as a rise of temperature of the bearings. This temperature inequality is then further smoothed out by thermal conduction into the surroundings of the bearings until finally a dead level is reached in which all large-scale motion has died out and all temperature differences have decayed to a final dead level.

This is typical; actual mechanical motion disappears and also as far as possible the potentiality for generating mechanical motion. This includes such things as the falling of weights to their lowest positions, the uncoiling of springs, or the running of possible chemical reactions. Along with the loss of actual or potential large-scale motion goes smoothing out of temperature inequalities. The extent and relative magnitude of these various smoothing-out processes may vary in almost any way from system to system, and depends on the detailed construction of the system. There is no reason why a weight originally on the floor should not be elevated to the ceiling by the uncoiling of a spring and maintained there by the action of a ratchet mechanism, nor is there any reason why the temperature differences which are generated by mechanical or chemical action should not be imprisoned by membranes of vanishingly small thermal conductivity and maintained for intervals of time which for all paper and pencil purposes may be put down as infinite. In many cases it is easy enough to see when these hindrances to a complete final smoothing out are operative, but in other cases it is impossible to be sure that the lowest potentialities have been realized, as in that of the test tube of subcooled water that suddenly changes to ice after ten days' quiescence, or the glycerine in the chemical laboratory at Berkeley that all froze solid after the importation of a nucleus of the solid from Oregon. There is no way whatever of being sure that any of the ordinary objects of daily life do not have other polymorphic forms into which they may sometime change spontaneously before our eyes.

Our command of the changes which spontaneously occur may seem restricted and unsatisfactory enough, but there is at least one sort of statement that can be positively made, namely that the direction of the sequence of changes which occurs is not capricious, but if a particular sequence starts or runs at all it will run in a unique direction: the sequence in a test tube of subcooled water which is in the act of turning itself into ice is always in the direction from liquid to solid, and never reverses itself from solid to liquid. We would like to be able to go farther in this statement and say that the direction of the changes which run spontaneously is always toward the attainment of the final "dead level." But there is a real problem to be solved before we can take this step, which otherwise might be reduced to a pure verbalism. For, when we reflect that changes spontaneously occur, such as the raising of a weight by the uncoiling of a spring or the discharge of a projectile by an exploding charge of powder, what meaning shall we give to "direction toward a dead level" independent of the direction which the change is actually observed to take? Most physicists would feel, I think, that there is something more than a verbalism in this situation, and it was the great contribution of the early thermodynamicians, Carnot, Kelvin, and Clausius, to disentangle the verbal from the physical. This disentanglement was affected by considering the system only in certain standard stock situations, of the greatest possible simplicity, and then establishing the sequence in which these stock situations do as a matter of fact occur. Thus if a system comprises a weight whose height is variable and other parts which are initially all at the same temperature, then a possible sequence for spontaneously occurring changes is for the weight to become lower and simultaneously a part of the system to rise in temperature, *there being no other change anywhere else in the system.* Whereas a sequence which never spontaneously occurs is for the weight to become higher and part of the system to become cooler than it was initially, again there being no other change anywhere in the system.

We now have the method for justifying our verbal impulse, and can say that what we mean by "in the direction toward a dead level" is the direction typified by the first sort of change, and conversely, what we mean by the "direction away from a dead level" is typified by the second sort of change.

A natural corollary of the proof that this verbalism can be justified is that it should be possible to get some sort of numerical measure of the degree of progress which a system has made in the direction toward its dead level. This also turns out to be possible and it is accomplished by the introduction of the concept of entropy and of the entropy function. The details of the way in which this was done by Kelvin or Clausius need not be reproduced here, but may be assumed sufficiently well known from the expositions in the textbooks. A preliminary to the introduction of entropy is the introduction of the notion of absolute temperature. As we have already seen, it is possible to define a temperature scale of sorts in terms of the properties of any arbitrary substance. The scales obtained in this way will of course be different for different substances. By a suitable distortion of these scales it is possible to obtain a scale which is the same no matter what substance affords the starting point. Many such "universal" scales are possible; probably the simplest of these is the "absolute" scale of Kelvin. The derivation of this scale is based on the generalization already mentioned, namely that there are certain sorts of change which do not occur spontaneously of themselves in closed systems, or which cannot be made to occur by the manipulation of an outside observer. As used in thermodynamics, these two methods of characterization are practically equivalent, for a process initiated by an unseen outside observer would be described as spontaneous, and on the other hand I shall not assume that I am any less potent than any other observer or manipulator. The situation is largely verbal. The manipulations of the observer, however, are restricted to those which involve only vanishing amounts of energy. Granted now that certain sorts of things cannot be made to happen, the notion of a heat engine with maximum efficiency naturally presents itself. The details of the analysis do not concern us; the general tactics of the analysis is merely to show that if engines could be constructed of greater than the maximum efficiency it would be possible to bring about the forbidden state of affairs.

REVERSIBILITY AND RESTORABILITY

The exigencies of the analysis and the proof lead to the introduction of a special kind of engine of maximum efficiency, namely the completely reversible engine. From this more or less accidental fact it has come about that reversible engines and reversible processes play a very important role in the conventional thermodynamic expositions. I would like to make the comparatively minor point that the emphasis on reversibility is somewhat misplaced. Instead of reversible processes one might better speak of processes of maximum efficiency. It is not the reversibility of the process that is of primary importance; the importance of reversibility arises because when we have reversibility we also have *recoverability*. It is the recoverability of the original situation that is important, not the detailed reversal of the steps which led to the original departure from the initial situation. That it is the recoverability which is important is evident at once from the formulation of the sort of thing which cannot be made to happen. This formulation states that certain sorts of simple change do not occur in isolated systems, all the other parts of the system (except for a weight or a heat reservoir) being restored to their initial state. During the intermediate operations the other parts of the system will in general deviate from their original condition. Unless these parts can be restored to their initial condition the formulation cannot be applied. Whence the importance of restorability in general. It is of course to be added that a formulation in terms of restorability would not have been a fruitful formulation unless the possibility of restoration to the original condition were so universal.

The argument that it should be possible in detail to replace strict reversibility by restorability may be very simply put. But we first digress for a moment to remark that although it is usual to think of purely mechanical systems as completely and necessarily reversible this is certainly not the case. It is possible to make any simple mechanical device, such as a flywheel in frictionless bearings, into a non-reversible one-way device by the simple addition of a ratchet arrangement which permits motion in one direction only. The mass of the ratchet mechanism may be vanishingly small, it may absorb no energy, and when the rotation is in the forward direction it may be absolutely without effect on the performance of the flywheel. The reason that this sort of thing is not usually considered is that the mathematics of a system of this sort would certainly be much more complicated than the mathematics of the conventional "mechanical" system. The mathematical approach to mechanics has become so important that we have unconsciously come to think of a mechanical system as defined in terms of its mathematics rather than in terms of its actual construction. Given now this possibility of making any mechanical device a one-way device, consider the engines used in the conventional proof of the maximum efficiency of reversible engines. If the ratchet device were placed on that engine which operates in the forward direction, nothing whatever observable would be altered. All the heat intakes and outputs and conversions into mechanical energy would be unaffected, and the efficiencies would also be unaffected, which we know are the thermodynamic efficiencies. But the conventional proof of what these efficiencies are now becomes impossible. The engine remains an engine of maximum efficiency, but it has become irreversible. Hence it must appear that it is the maximum efficiency which is important and not the reversibility. Incidentally, this suggests that it should be possible to find some other method of establishing what the maximum efficiency is (which amounts to defining the absolute temperature) than by the use of the conventional completely reversible cycle. This I believe can be done rather simply, but I do not want to go into such technical details here.

It appears to be possible to go still further in the same direction and get along with a different and perhaps weaker requirement than recoverability. The only requirement for a point energy function is that the final states should be attainable from the initial states by a number of paths, continuously deformable into each other. It is not at all necessary that it should be possible to reach the initial states from the final states by any process whatever. If, for instance, in a gravitational field, some sort of ratchet mechanism made impossible any motion with a component toward the surface of the earth, while freely admitting all motions away from the surface as now, all possible motions would be describable with the same potential function as now. The physics of a world which did not have recoverability would be strange, for no experiment could ever be repeated with the same piece of matter, and most engines would become impossible. The operations of an experiment could be repeated, however, with new pieces of matter with a past history identically the same as the matter previously used, and the same mathematical functions as now could be used to describe the results. There is no point in elaborating this idea further because such a world is so far from the world of experience, but it does at least suggest that our mathematics does not fit our experience as exactly as we are sometimes inclined to suppose.

The meaning of "reversible" is usually assumed to be so obvious that it can be passed over without further analysis; nevertheless there are further points which I believe require comment. A process may be said to be reversed when its sequence of configurations is reversed in time. The possibility of ambiguity arises in connection with the meaning of "configuration." In the case of simple thermodynamic systems there is not much difficulty, and "configuration" is pretty much the same as thermodynamic "state." A thermodynamic process is reversed when the sequence of "states" of the system is reversed in time. This means in the most elementary situation that the sequence of temperatures and pressures at every point of the body is reversed. The difficulty arises with regard to what shall properly be called the state parameters in more complicated systems. In mechanical systems, should the *velocity* of the elements of the system be considered as one of the parameters of state? From one point of view the velocities certainly should be considered among the parameters of state, because the kinetic energy of the system, and therefore its total energy, depends on the velocities. But on the other hand, when we reverse the sequence in time of the positions in space of a mechanical system, we reverse the direction of every velocity; it is only the sequence of absolute values of velocities which is reversed, not the sequence of velocities themselves. The dilemma becomes even more apparent in an electromagnetic system. There would not seem to be much question but that the electric and magnetic vectors should be considered among the parameters which fix the state of the system; certainly the internal energy depends on them. Yet if we reverse the sequence in time of the positions of the electrical particles we have thereby reversed their velocities and thereby reversed the *direction* in space of all the resulting magnetic forces, not merely the *sequence* in time of the magnetic forces. The conclusion would seem to be that reversibility should not be defined in terms of a reversal of sequence of *all* the parameters of state, but that only a restricted group of these parameters is to be included in the requirement of reversibility. This affords a means of dividing the parameters of state into two groups, one analogous to the coördinates of position, and the other to generalized velocities.

No process in which there is heat transfer on a temperature gradient can be strictly reversible. Such transfers can be handled in arguments demanding reversibility by various limiting processes, involving differentials of various orders.

ENTROPY

Given now the absolute scale of temperature, it is possible to define in terms of it a new function of state of a body in the same way that the energy function was introduced with the aid of the first law. This new function, the entropy, is in the first instance defined by means of reversible processes, just as the energy was in the first instance defined for processes for which dW and dQ had a meaning. The formal definition $S(A,B) = \int_A^B \frac{dQ}{T}$ exhibits entropy as depending on an initial and a final state, that is a state-couple, just like the energy. Like energy, entropy is in the first instance a measure of something that happens when one state is transformed into another. The second law then shows that for reversible processes the integral is independent of the intermediate details and depends only on the initial and final states. This situation is verbalized in the same way as that for energy by talking about the *difference* of entropy between two states, with the almost unavoidable implication that it must mean something "physical" to talk about the entropy of a single state. Mathematics, which as Willard Gibbs emphasized is a language, lends itself to this verbalizing because of the particular properties given to entropy by the definition. For it follows from the definition, as it does also for energy, that $S(A,B) + S(B,C) = S(A,C)$. The solution of this functional equation is $S(A,B) = S(A) - S(B)$. The situation is dealt with most simply from the mathematical point of view by selecting any convenient state of the body as the arbitrary zero state from which to calculate changes of entropy, and assigning an arbitrary value to the "entropy" of this state. The change of entropy in passing from this state to any other which can be reached from it by a reversible process is then defined by the integral $\int \frac{dQ}{T}$. But once the state function has been found, it has a wider significance than application merely to the reversible processes which formally generated it, analogously to energy. This wider significance is to be found in the behavior of entropy when irreversible processes occur.

The change of entropy of a heterogeneous system from a given initial point is defined as the sum of the changes of entropy of each of its elements, taken small enough so that they may each be regarded as homogeneous. We assume for the present that each of the homogeneous elements may be connected by reversible processes to their initial states, so that the changes of entropy of the elements are themselves defined. By means of the definition it is now possible to assign a value to the change of entropy of the total contents of an isolated enclosure after some process has taken place, whether or not the process is reversible. But in order that the entropy of a heterogeneous system so defined be profitable, certain conditions must be satisfied. If every process taking place within the enclosure is "reversible," then we require that the net change of entropy of the entire contents after the process be zero. On the other hand, if any portion of the process is not reversible, that is, not of maximum efficiency, then the net change of entropy must be an increase. It is the *experimental* fact that these conditions can be met that justifies the definition of the entropy of a heterogeneous system as the sum of the entropy of its elements regarded as homogeneous.

The business of analyzing a heterogeneous system into small elements each of which is treated as homogeneous requires, I think, a little more consideration than is usually given it. Such an analysis is adequate only for certain aspects of the behavior of the system. It is adequate, for example, for the energy. In simple mechanical systems the parameters which are assumed uniform throughout an element are pressure, temperature, and velocity, and an energy obtained by integrating the proper function of these parameters over the body satisfies the condition that it remain constant for every possible combination occurring inside an isolated enclosure. It is similarly adequate for entropy, because it is possible to satisfy the conditions just mentioned. But such an analysis is not adequate to give the time rate of change of entropy, for example; to fix this in a simple mechanical system an additional parameter is needed at every point, namely the temperature gradient.

Having now our definition of the entropy of a heterogeneous system, we may at first merely as a matter of further definition take the magnitude of the increase of entropy during the process as a measure of its amount of "irreversibility," the sort of thing that we were looking for. Examination shows that this is indeed a felicitous definition, for it lends itself to the sort of verbalizing that is natural. For instance, the transfer of heat by conduction from one body to another is, in the limit, when the temperature of the two bodies is the same, a reversible process, or one of maximum efficiency. It is also one for which there is no net change of entropy. But if the temperature of the two bodies is different, the process is "irreversible" (heat cannot be made to pass by conduction up a temperature gradient), and the entropy increases. Furthermore, the increase of entropy is greater for the same amount of heat transferred the greater the temperature difference, and this is what we would like to be able to say in view of the fact that the entropy increase is zero when the temperature difference is zero. Or again, the amount of work that a system can deliver is evidently connected with its progress toward the "dead level," for when the dead level is reached, no work can be extracted. Detailed examination shows that when entropy-increasing processes take place in a system its capacity to deliver mechanical work decreases, and the amount of entropy increase is proportional to the loss of capacity to deliver work. This may be expressed by saying that the increase of entropy is a measure of the loss of "availability" of the energy of the system. This again is the way that we would like to be able to talk about the situation.

Given now the entropy of a homogeneous body as a single-valued function of its parameters of state for all those states which can be reached by a reversible process from some state arbitrarily chosen as the initial state, the entropy of which may be set equal to zero by convention without loss, the question arises as to the entropy of states which cannot be reached by reversible processes from the initial state. There are such states, of course, as shown by the change from graphite to diamond. The analogous problem with respect to energy was solved by putting the recalcitrant piece of matter in an enclosure with other sorts of matter whose energy properties had already been completely mapped out, allowing the process to run, and setting the change of energy of the recalcitrant piece equal to the negative of the determinable change of energy of the rest of the matter. A similar procedure may be adopted here. Diamond may be turned to graphite in an enclosure in the presence of other matter, and the change of entropy of this other matter may be determined after the process has been completed. The difference of entropy between graphite and diamond must now be such that the net change of entropy is an increase. But here we are much worse off than we were with the energy, because the process has given us merely an inequality, instead of an equality from which in the case of energy we were at once able to get a unique answer. A single experiment is evidently insufficient to fix the unknown change of entropy.

We might of course perform a great many different sorts of experiment, and attempt to get some satisfactory answer from all these experiments together. It is at any rate a necessary condition that no one of this multiplicity of experiments shall give a negative change of entropy. But this condition is evidently not enough. If we could arrange the various experiments in some sort of a hierarchy according to their "degree of irreversibility" we would be on the way to our solution. But this would be of prohibitive difficulty in practice. It would involve in the first place being able to at least approach to the limiting case where there is zero irreversibility, which is nothing but the reversible method of transfer which by hypothesis we could not accomplish. Furthermore, to assign a numerical value to the degree of irreversibility of any of the alternative processes would demand a knowledge of the intimate details of the process quite beyond our powers. In fact, in many cases to say that we cannot give the details of a process (which means to give so much data that whenever these data are repeated the whole future measurable course of the system is repeated) is equivalent to saying that the process is irreversible.

We have, then, a method of dealing "in principle" with the entropy of states of matter not reachable reversibly (namely that the net increase of entropy must always be positive for all possible combinations in which the state can be reached), but this method in practice reduces to something of very little value because of the difficulty of realizing the range of conditions demanded. One might perhaps try to give a unique value to the entropy through the "third" law, but examination will show that this is not of much help. The value of the entropy obtained by applying the third law is checked, and therefore has its meaning, only by describing some reversible transformation. Furthermore in practice the application of the third law is made uncertain by the possibility that there may be finite transformations at temperatures closer to the absolute zero than have yet been explored. If there is no reversible method for getting from one state to the other, I believe that the only thermodynamic method of giving meaning to the entropy must follow the general lines of the last paragraph in spite of the difficulties. Although this thermodynamic method of extending the meaning of entropy is mostly formal and verbal, nevertheless the possibility should be kept in mind in our pencil and pen manipulations. If the method is not "thermodynamic," then we have really altered the meaning of "entropy."

One may perhaps envisage the possibility of extending this sort of procedure to obtain the specific values of the entropy of any special substance in terms of its parameters of state without knowing the entropy of *any* comparison substance or knowing how to make the special substance take part in reversible processes. Such a procedure would be analogous to the most general procedure suggested for the energy. Given, in the first place, that an entropy function of the parameters "exists": then it is conceivable that the precise value of this function, except for its known indetermination by an arbitrary function, both additive and multiplicative, might be determined by the condition that the total entropy of the contents of an isolated enclosure always increases for all conceivable combinations of the given material with other materials inside the enclosure, for all conceivable processes, and for all conceivable initial conditions, whenever there is any irreversible aspect to the process. The initial and final conditions must of course be subject to the condition that they can be broken up into elements homogeneous in the selected parameters. For intermediate states in the processes, when there are no macroscopic parameters with which the system can be characterized, the entropy is undefined. It would seem that the order of infinity of combinations of all conceivable materials in all possible relative amounts with all conceivable processes and all conceivable initial conditions must be higher than the order of the infinity of the functions of the parameters to be determined. This should, then, be a theoretically possible method of determining the entropy function. I suspect, however, that no one would seriously propose it as a laboratory method. Even if it could be made to work its value would be almost purely descriptive, for we could never be sure that we had exhausted all possible combinations until we had exhausted experience.

HOW SHALL WE HANDLE IRREVERSIBILITY?

The difficulty we have had in giving meaning to entropy for essentially irreversible processes runs through and colors all thermodynamics. It is almost always emphasized that thermodynamics is concerned with reversible processes and equilibrium states, and that it can have nothing to do with irreversible processes or systems out of equilibrium in which changes are progressing at a finite rate. The reason for the importance of equilibrium states is obvious enough when one reflects that temperature itself is defined in terms of equilibrium states. But the admission of general impotence in the presence of irreversible processes appears on reflection to be a surprising thing. Physics does not usually adopt such an attitude of defeatism. Of course this may be made a matter of words if one chooses, and one can say that thermodynamics by definition deals only with equilibrium states. But this verbalism gets nowhere; *physics* is not thereby absolved from dealing with irreversible processes, or we from trying to understand a little better this anomalous situation.

It seems to me that there are two essentially different aspects to this question of irreversibility. In the first place, many processes are of great complexity, of a complexity so great that we cannot adequately describe them by means of the instrumental operations that we can perform. The meaning of "adequate" description is, as always, such that whenever the terms of the description are again applicable to a system, the subsequent evolution of that system repeats its previous course, the description of the environment also being the same. Systems which are so complex that adequate description is impossible are thus not "causal" systems. It is generally accepted that such systems are not amenable to scientific treatment, disregarding quantum phenomena for the present. An example of such a system is a hydrodynamic system breaking into turbulent motion. The "irreversibility" of many processes arises because they are so complex that adequate description is impossible, for the practical attitude of thermodynamics toward such complicated processes will be merely refusal to have anything to do with them, which is also its attitude toward irreversible processes.

On the other hand, there are a number of irreversible processes which are so simple that they can be adequately described in terms of such large-scale instrumental operations as are presupposed in thermodynamics. An example of such a process is conduction of heat down a temperature gradient. Other such processes are: development of Joulean heat when an electrical current flows against Ohmic resistance; diffusion of solid, liquid, or gas down a concentration gradient; laminar flow of a viscous liquid; and perhaps certain aspects of the crystallization of a subcooled liquid. All these are simple processes and are adequately described in terms of a few variables. For instance, when we have an electrical current of given intensity flowing in a conductor of given resistance, all the other measurable aspects of the phenomenon repeat.

A more detailed study and classification of irreversible processes than any yet attempted would doubtless be rewarding. Thus besides those processes in which "irreversibility" occurs because of the impossibility of adequate description, there are processes which are spontaneously initiated so deep down in the structure of things that any detailed study by instruments of the thermodynamic scale is forever ruled out. Such a process is an atomic disintegration — indeed here it is a cardinal principle of wave mechanics that adequate description of such a disintegration is meaningless. Somewhat analogous to such disintegrations in general effects are the occurrences that initiate the crystallization of a subcooled liquid, or of a supersaturated solution, or the running of some chemical reactions. The formation of the nucleus that initiates these things is just as capricious from the point of view of the instruments of thermodynamics as an atomic disintegration, so that a system containing a sub-cooled liquid must be called a non-causal system. But even here I question whether there is any close connection between capriciousness and irreversibility. A single nucleus can set off the crystallization of the whole mass, but the nucleus is so small that the process of its formation can hardly make any appreciable contribution to the final increase of entropy of the whole mass. The major part of the crystallization process consists of the advance of the surface of the solid into the liquid. This advance is a complicated thing — there is latent heat to be got rid of and there will be a complex set of convection currents, so that the large-scale process may well appear non-causal because of the difficulty of sufficiently characterizing these fine details. But that part of the crystallization process which occurs at the very surface where the atoms of solid are being laid down out of the liquid appears to proceed with all the inevitability and reproducibility of a flow of heat down a temperature gradient, and this aspect of the phenomena should be amenable to treatment.

There are two criteria which are popularly considered to be certain marks of irreversibility, namely spontaneous occurrence of a process, and the running of the process at a finite rate. I do not believe, however, that either of these is a sufficient criterion, for examples can be set up in the fields of mechanics or radiation, as will appear later, that violate them. It seems to me that such processes should be capable of treatment by methods thermodynamic in spirit even if one is unwilling to use the word "thermodynamic" in its strict formal sense.

I have made the beginning of an attempt to treat certain irreversible phenomena by methods thermodynamic in spirit. The fundamental postulate is that when any well-defined irreversible process occurs it is accompanied by its characteristic increase of entropy. The location of this increase of entropy is determined by the details of the set-up, but the universe as a whole bears the indelible trace of the occurrence of such an irreversible process in a characteristic increase of entropy somewhere. This postulate goes beyond "classical" thermodynamics, which would merely say that in the limit when the process becomes reversible the increase of entropy goes to zero, and that when there is an irreversible process there must somewhere be *some* increase of entropy without specifying its precise amount. It is probably too early to feel convinced that this new formulation is applicable in every situation, but I have at least obtained results which agree with experiment in the known cases, and are in contradiction with experiment in no known case. Furthermore, sometimes the results cannot be obtained by the conventional methods, so that something new is involved.

I shall not go into the details of these applications here, which will be found in my book on *The Thermodynamics of the Electrical Properties of Metals.* It will pay, however, to give an outline of perhaps the most important application. Thermoelectrical phenomena in metals should be completely amenable to attack by thermodynamic methods, since the phenomena can be adequately described in terms of a few simple measurements with large-scale thermal and mechanical ("mechanical" generalized as always to include electrical) instruments. One of the first applications of thermodynamics made by Kelvin was to these phenomena. Kelvin's attack was not by purely thermodynamic methods, for he introduced an additional hypothesis, of great verbal plausibility, but one which did not fit into the known scheme of things and which logically had no other justification than that it gave the right results. Kelvin always was dissatisfied with the situation, and one of his last papers was an unsuccessful attempt to get the same result rigorously. The difficulty in the situation arises because there are *two* irreversible processes which accompany the thermoelectric process, namely thermal conduction and the development of Joulean heat. Kelvin postulated that the irreversible processes could be completely disregarded and thermodynamics applied to the thermoelectric phenomena alone, which were postulated to be reversible, just as if the irreversible processes were absent. A similar sort of thing had been done often enough in other situations, but there was this vital difference: in the other situations the irreversible aspects could be made to vanish in comparison with the reversible aspects by a suitable choice of the dimensions of the apparatus or the time of the experiment, or of some other parameter. In the thermoelectric case, however, the two irreversible phenomena are so tied together that if one is made to vanish the other becomes large. The two together can be made only to have a minimum effect, which is so large compared with the reversible effects of the thermoelectric action that a strict application of thermodynamics, as was made by Boltzmann, gives an inequality which fails by a factor of one thousand to touch the actual physical situation.

If, however, instead of trying to suppress or to postulate away the irreversible aspects so that conventional thermodynamics can be applied, one deals with the irreversible phenomena *in situ* as they occur alongside the reversible phenomenon by means of the postulate that the occurrence of the irreversible effects is accompanied by a certain definite increase of entropy, one comes out with precisely the relations of Kelvin, which are checked by experiment. The same sort of analysis can also be applied to the various transverse thermal and electrical effects, of which the Hall effect is the best known, and yields relations which appear to check with experiment, and which had not been previously deduced. In my book will be found a few other applications. One interesting result is the concept of a "thermo-motive" force, the analogue of the "electro-motive" force of electrodynamics, which under proper conditions may maintain a temperature gradient in a substance without any corresponding flow of heat. A consequence of the possibility of "thermo-motive" forces is that the irreversibility which we ordinarily associate with a flow of heat does not arise primarily from the *flow* but from the fact that the flow is usually the result of a *conduction* mechanism. It is quite possible to have a finite and *reversible* flow of heat if the flow is maintained by the action of a thermo-motive force. It should be possible to extend the general method to phenomena of diffusion, etc., although I have not seriously attempted this. I believe that when the method has been sufficiently applied it will be found that there are no sorts of essential irreversibility which do not come under this scheme: apparent exceptions arise only when we have reproducible irreversibility occurring in circumstances so complicated and of such a small scale that adequate description becomes impossible. An essential preliminary to a complete working out of the idea is an exhaustive listing of all the detailed forms which essential irreversibility may take, just as the energy concept acquires its validity only by an exhaustive cataloguing of all the different kinds of energy transformation.

This suggested method of treating certain irreversible processes seems to me to be entirely in the spirit of the classical thermodynamics, only it carries the idea a little further and gives it quantitative form. The classical picture was that any irreversible process leaves its indelible imprint on the "universe" ("universe" here means merely any isolated system); certain measurements made on the universe after the irreversible occurrence are different in a specifiable direction from the same measurements made before the occurrence. To this has now been added the notion that there are certain characteristic irreversible processes which may occur independently or together but are such that their effects are assignable and additive as far as their permanent traces are concerned.

THE FLOW OF ENTROPY

It should be possible to give a more detailed account of what the permanent trace left by irreversibility on the countenance of the universe consists in. Statistical mechanics does give such a detailed picture, and we shall return to this picture presently. But it should be possible to give a less detailed account in the spirit of thermodynamics, that is an account in terms of the results of measurements by the same instruments which give the thermodynamic parameters of state. This of course has already been done in a sense when the particular function of the parameters of state has been discovered for every substance that is its entropy function. But I think we would like to be able to do more; certain questions that we can ask should have an answer. For instance, is there not some connection between the place where the final change of entropy resides and the place where the irreversible process occurred? What is the pattern by which entropy changes locally? does it make sense to talk about a flow of entropy? This is an example of what happens when we carry familiar patterns of verbalizing into new situations; we have at first no guarantee at all that the old verbalizings will even have meaning, but we do at least in this way provide ourselves with a program of exploration.

Consideration of various simple cases will indicate the answers that we must give to some of these questions. Suppose, for example, that we have two reservoirs at different temperatures and allow heat to pass by conduction from the hotter to the colder through a connecting bridge of conducting material. The mass of the reservoirs may be taken to be large compared with that of the bridge, and the conduction may be allowed to run for a long time. The mass of the reservoirs is large enough so that in this time there is no appreciable change of temperature of the reservoirs, and the material of the reservoirs is supposed to be so good a conductor that the temperature of each reservoir is uniform at all stages — rather a formidable combination of properties to realize in the laboratory. We then know that the entropy of the hotter reservoir decreases by the amount Q/τ_1 where Q is the amount of heat which has left it and τ_1 is its temperature. The entropy of the cooler reservoir at the same time increases by the amount Q/τ_2, where τ_2 is the temperature of the cooler reservoir. Since τ_2 is less than τ_1, the increase is greater numerically than the decrease, so that on balance the entropy of the universe has increased during the process, as it should. The change of entropy of the conducting bar is negligible, both because its mass is negligible and because it is in a steady state so that there is no change of its parameters of state as the process runs. This at first blush might appear paradoxical, because it is just in the bridge that the irreversible process of heat conduction runs. We conclude in general, then, that the final location of the entropy change may be different from the location of the irreversible process which generated it. It would therefore appear that we are forced to "say" that entropy flows. Furthermore, there need be no irreversible process occurring at the place where the "flow" occurs. This can be seen in the simple example by considering that the increased entropy of the cooler reservoir is uniformly distributed throughout it and that no irreversible process occurs in the reservoir. It is the generation of entropy that is the result of irreversibility; once generated, entropy may be transported from place to place without necessarily implying any further irreversibility.

It is now pretty obvious how we should verbalize the situation. We can (or perhaps *must*) say:

> (Total entropy leaving a closed region) — (Total entropy entering the region) = (Total entropy created within the region) – (Increase of entropy localized in region)

Or, another way:

> (Net entropy leaving a close region) = (Entropy created within the region) – (Increase of entropy localized in region)

Or, this may be expressed in terms of the entropy flow vector, $(\vec{S})$, by the equation:

$$\text{Div } (\vec{S}) = -\frac{\partial S}{\partial t} + C$$

Here S is the entropy function per unit volume of the given material in terms of its state parameters, and C is the rate of creation of entropy per unit volume. C is at first a pure convention, defined by the equation, but if we could have C a measure of the degree of irreversibility of the process occurring at the point we would feel that we were getting some physical content into the equation. Verbally, the situation is entirely similar to that which we have already examined with respect to energy. There is this difference that here we have some indication of an independent instrumental procedure which may remove the purely conventional nature of "creation."

It is obvious that our equation is applicable to the case we have just considered of conduction from one reservoir to another. Take the bridge of unit cross section and let the heat current be $\vec{q}$ per unit time across this section, in the direction of the axis of the bridge. Then $\vec{S} = \frac{\vec{q}}{\tau}$ where τ is the absolute temperature at a point. The amount of increase of entropy when a quantity of heat Q passes by conduction from τ to $\tau - \Delta\tau$ is $Q\Delta\tau/\tau^2$, so that in our example $C = -q \cdot \text{Grad}\ \tau/\tau^2$. Furthermore, since the state of the bridge is everywhere steady, $\delta S/\delta t = 0$. Substitution of these special values at once checks the relation on p. 142. This sort of thing appears to be possible in general whenever there is heat conduction.

The same sort of consideration applies here that we have already entertained with regard to energy. As originally written, our equation defines a purely conventional flow of entropy, because a vector may always be found such that its divergence has an assigned value. But what we have done here has real "physical" significance, for we have found an independent instrumental meaning for all the terms of the equation, and then it may be checked by experiment that these independently found terms satisfy the given relation of equality.

This treatment demands several comments. In the first place it is to be emphasized that the quantities which we have used in our formulation all satisfy the criteria which we have previously imposed for "physical reality" in the sense that all of them can be found in terms of operations with instruments made at the point in question. Thus the entropy flow vector at a point is found in terms of a measurement of the temperature at the point and the heat flow vector at the point. This latter is indirectly determined in terms of a measurement of the temperature gradient at the point (surely as legitimate a "physical" thing as the velocity of a particle at an instant of time), and the thermal conductivity of the material at the point. (The fact that sometimes, as in the interior of either reservoir, we talk about a heat flow with no temperature gradient need occasion no real difficulty, but may be dealt with in terms of temperature infinitesimals of higher orders.) The rate of creation of entropy C is determined from the same measurements that were used in getting $\vec{S}$. Finally, $\delta S/\delta t$ is obtained from a measurement of the state parameters and their changes with time, combined by paper and pencil manipulations with the entropy function which has been determined by previous rehearsals.

The complete physical significance of what we have done in connection with this problem of the flow of entropy associated with thermal conduction is not merely that we have been able to remove the conventional character from our verbalizations by finding a possible set of instrumentalizations. In any specific situation it should always be possible to instrumentalize any specific conventional verbalization. The wider significance is that the things which we do to give meaning to "flux of entropy" and the other terms of our formulation are independent of the particular system and the context in which we may for the moment be interested. Our formulation covers an infinity of high order of physical systems in terms that can be taken out of context. It is true that our rules may have demanded previous rehearsal to assign such things as the entropy function of specific materials, but these rehearsals are for only a few typical situations, and when once made enable us to deal without rehearsal with infinitely more numerous other situations. It is a discovery to find formulations to which one is driven by verbal compulsion which have this wider significance; a large part of science is concerned with the search for and the discovery of such formulations. The question "why" such formulations are possible at all is a tempting one but probably meaningless. Such meaning as it has would involve a detailed study of the evolution of language.

We have seen that the "physical reality" that we have been able to give to the flux of entropy at a point and other similar quantities means that the instrumental operations involved are those made at the point and at the time in question. We would expect, in the most general situation and in the absence of other information, that the instrumental operations which give meaning to physical quantities would involve operations at other places and over intervals of time, that is, should be functions of the surroundings. The fact that we find that we can so often dispense with any such reference to the surroundings is doubtless connected with the other experimental fact that our rules can often be so formulated that they are always applicable, no matter what the context or surroundings. Mathematically, a rule the same for all possible contexts or surroundings is one independent of the surroundings.

Again we have a situation which from one point of view is pretty hopelessly verbal. The thing of physical significance is that as a matter of fact we are able to formulate rules of operation which are independent of their physical surroundings and which at the same time do fit the empty molds of verbal compulsion.

The equation which we have written above for flow of entropy is not often used, nor is this subject often discussed. However, something quite similar will be found in some of the writings of Ehrenfest. There is, I think, no reason to question its correctness. Doubtless it is capable of many more applications and of giving new results. In the general case the term for the rate of creation of entropy must be modified if there are other irreversible processes occurring along with heat conduction. Perhaps the simplest of these is development of Joulean heat; the additional term to be used for this is obvious. Then, too, in more complicated cases the expression for the vector flow of entropy has to be modified. For example, if an electrical current is flowing, we have to "say" that there is a certain transport of entropy by the electrical current. The precise expression is not difficult to write, and involves the Thomson coefficient of the conductor. An application to thermoelectric systems will be found in my book. Or, if we are dealing with non-isotropic substances, the direction of the flow of entropy is not the same as the direction of heat flow.

Having found that it is possible and probably profitable to talk about a flow of entropy, we may raise the further question, as we did in the case of energy, as to what extent we may apply other verbal habits in the situation. Does it make sense to talk about a *velocity* of flow of entropy, and if it does, what is the density of entropy to which we are thereby led by the condition that the product of velocity and density is equal to the rate of flow? Of course, if we could begin at the other end the answer would be simple enough, because if we had some natural value for the density then the velocity would be given by the same relation. This procedure was possible for the electromagnetic field because there was a natural value for the density of electromagnetic energy to which we had been led by other considerations. But in the case of entropy there is no such natural density, because there is no natural zero from which to figure entropy, at least as long as we stick to conventional thermodynamic operations that can be performed in the laboratory. With regard to any possible *velocity* of flow, the same conditions are to be placed on this velocity as on the velocity of any material thing. In particular, the velocity must be additively affected by describing the velocity in a moving frame of reference. When one considers that there are thermodynamic systems which are composed partly of ordinary matter and partly of radiation trapped in the space in cavities in the matter, and that this additive term would have to affect equally the material and the non-material parts of the system, I question whether a consistent solution is possible. However, these considerations are not final, and at present all I can say is that I have not thought enough about the situation to be sure of the answer. It is at any rate certain that the analysis of entropy flow into a velocity and a density factor is not usually made, so that the probability is that even if such an analysis is possible, the necessity for it can be side-stepped.

THE ENTROPY OF THE UNIVERSE AND THE ENTROPY ASSOCIATED WITH RADIATION

Another question to which we are led naturally by letting our ordinary verbal impulses have free play is as to the meaning of the "entropy of the entire universe." Of course our original definitions have no application, for they were limited to reversible processes in isolated systems, and the universe is not isolated, nor are its processes reversible. But we found that in the case of energy we could formally extend the original definition to include the "entire universe," and it is quite conceivable that the same sort of thing may be done with entropy. At any rate, it is very common for physicists to talk about the entropy of the entire universe, and one of the favorite formulations of the second law is that the entropy of the entire universe is always increasing and is striving toward a maximum. The only formal justification would be that the whole universe is merely a special case of an isolated system. To justify this one would have to think of the universe as a finite nucleus of matter, surrounded by an expanding aura of radiation in empty space, with a region of radiationless and fieldless space always surrounding the occupied part. Recent cosmological speculations remove some of the inevitable naturalness from this picture, to say the least. Given the picture, one has to have some way of calculating the entropy associated with the material parts of the system and also with the field of surrounding radiation. The material part may be dealt with readily enough. We suppose that we know the entropy function of all the particular kinds of matter as a function of their state parameters, and that we know all the necessary state parameters.

The entropy of all the matter is then merely the sum of the entropy of all the elements, exactly as it was in the case of the energy. Because the matter in the universe is continually losing thermal energy by radiation, it follows that the entropy of the material part of the universe is continually *decreasing*. The net increase of the total entropy must therefore arise from a more than compensating increase contributed by the radiational field. One can at once say that the reason for this is obviously the ever expanding volume occupied by the radiational field. The parts of the field already occupied may be assumed to be in an approximately steady state so that they are making no contribution to the *change* of entropy of the universe, which is what we are really concerned with here, although we may have been careless in our expression. To carry through this suggestion in detail demands that we be able to write down the expression for the entropy density in an expanding radiational field. There are difficulties in doing this. It is true that Planck has given expressions for the entropy of radiation, but these apply to "black" radiation. Black radiation is characterized by perfect randomness; this involves a definite spectral distribution of energy and an indiscriminate mixture of all directions of polarization and propagation. Now although the stellar radiation from the universe might be admitted to have a spectral distribution sufficiently close to that of a black body for our purposes, it certainly does not satisfy the condition with regard to direction of propagation, for the radiation from the universe is unidirectional, continually away from the occupied part into the as yet unreached depths of empty space. This makes an enormous difference, as one can see by a simple argument. Idealize the universe as a sphere at uniform temperature which has been prevented from radiating since its creation by an impervious shield. Remove this shield, allowing the sphere to radiate a finite "gulp" of radiation, and replace the shield. The entropy of the sphere has decreased, but to compensate there is entropy traveling in the gulp of radiation. Next catch this gulp of radiation on a perfectly reflecting spherical mirror placed at the confines of the "universe," reversing the direction of radiation. The radiation eventually returns to the sphere, is there absorbed, and the sphere in the end resumes its initial temperature. The initial entropy of the sphere is therefore also resumed. There is no other place where there may be any entropy, so that the net result of the emission and reabsorption of the gulp must be that the total entropy of the universe is unchanged. But since the entropy of a closed system can never decrease, this means that the total entropy was always the same. The entropy associated with the gulp of radiation while it was traveling through space was exactly equal to the entropy lost by the sphere. Furthermore, since the occurrence of an irreversible process leaves its indelible impress in a permanent increase of entropy, it must be that no irreversible or entropy-increasing processes occurred. That is, neither the emission of radiation by the sphere in the first place nor its reabsorption later is to be described as an irreversible process. This I believe is quite different from what would be the snap judgment of many physicists. It is evidently the unidirectional character of the radiation that distinguishes this special situation. Although the unidirectional emission of radiation is a process which occurs spontaneously and at a finite rate it is not an irreversible or entropy-increasing process, as is the conduction of heat down a temperature gradient, a process which to our first crude physical intuition might "feel" to be the same. It is not uncommon to find in thermodynamic discussions the statement that any process which naturally occurs at a finite rate is necessarily an irreversible or entropy-increasing process. This generalization is obviously too sweeping; it also fails in simple mechanical situations, as already suggested.

It is a problem to find the best way to verbalize this situation. Shall the emission of a unidirectional gulp of radiation be described as *heat* emitted by the body or as mechanical work done by it? The usual method of description would call this radiant *heat*, but the perfectly reversible character of the act would seem more in keeping with what we would like to call mechanical work. I think if the radiated energy were monochromatic one need have no hesitation in referring to the radiated energy in terms of mechanical work, certainly one would have no hesitation in the limiting case of long radio waves.

Here then is a case where it is not immediately obvious how one should analyze an energy into a mechanical and a thermal part. A full answer to the question would demand that we know to what extent the unidirectional gulp of radiation has a characteristic temperature. This might be defined as the temperature of the emitting body, which before the act of emission certainly had a characteristic temperature to which it was forced by the physical boundary conditions. But if it is going to be profitable to define the temperature of the gulp as the temperature of the emitting body, then we must be able to show that all the measurable properties of the gulp are always the same when the temperature of the body is the same, no matter what the material or other physical properties of the emitter. It is by no means obvious that such would be the case; in fact one would almost be willing to bet that there would be at least part of the radiated gulp which would vary with the surface properties of the emitting body. This part of the radiation should almost certainly be described in mechanical terms.

A full study of the emitted radiation must involve a specification of the instruments to be used in its study. An analysis of the possibilities would be complicated. A full statistical analysis assumes the possibility of giving in microscopic detail at every point of space the local electric and magnetic vectors. Such a complete analysis is of course a paper and pencil analysis. Thermodynamics demands only an analysis with large-scale instruments possible to approximate in the laboratory. The results of measurements with such large-scale instruments would be used in determining what the thermodynamic parameters of state are. It is, I think, possible to imagine the construction of a number of large-scale instruments different from the conventional ones with which the properties of a gulp of radiation might be explored. For instance, it should be possible to construct a large-scale instrument which should analyze the directional character of the radiation, whether it was indiscriminately in all directions or what fraction of it was in a specified direction. The results of measurement with such an instrument would determine one of the parameters of state that must be specified to adequately describe the gulp, and this parameter would have to be numbered among the mechanical parameters.

The situation presented by radiation is therefore evidently complex. I shall not attempt a more detailed analysis. But the conclusion is to be kept in mind that in radiation problems the analysis of the action into a dW part and a dQ part may not be perfectly straightforward. With regard to the entropy of the "universe" it is evident that the localization and the details of the entropy-increasing process are not what might be assumed at a naïve first glance. Of course one may repudiate the entire picture at the basis of the argument above, and say that because of the unwelcome conclusions to which one has been led the fundamental assumption of an open finite universe must be discarded in favor of a finite bounded curved universe. This is certainly a possibility, but I do not believe it is the one to which we are impelled of necessity. The entropy-increasing processes of the universe are to be sought in the interior of the matter of which it is composed. Temperature inequalities result in the first place in thermal conduction from hotter to colder, always an entropy-increasing process. Furthermore, there is under such conditions radiation between hot and cold. The entropy increase arising from this process, that is, exchange of radiation between bodies at different temperatures, is not to be sought in the initial act of absorption, which may be non-entropy-increasing, but is to be found after the initial absorption in the spreading out of the spectrum of the absorbed energy from the distribution characteristic of the higher temperature of its source to the distribution characteristic of the lower temperature of the sink.

STATISTICAL ASPECTS OF ENTROPY

We have hitherto been concerned almost entirely with the thermodynamic aspects of entropy. The interest of physicists has, however, been much more directed to the statistical aspects, and the reason is doubtless that it appears to be possible in this way to find the answer to the many puzzling questions that present themselves with regard to the entire entropy situation. Yet paradoxically a good deal of the activity of physicists in this field has been directed to solving a difficulty presented just by the statistical picture. Any purely mechanical system, such as is assumed in the statistical picture, is perfectly reversible, or better, as already expounded, non-degenerative. At any rate, any purely mechanical system can by proper manipulation be brought back to its initial configuration. How, then, understand the occurrence of essentially irreversible processes, processes which leave their indelible trace which can never be wiped out by any method whatsoever? A more cynical or a more robust appraisal of the significance of this difficulty might have led to questioning the legitimacy of the fundamental operation of the statistical picture, namely extrapolation of the assumptions of classical mechanics into the paper and pencil domain far beyond verification with actual instruments. But when statistical mechanics was first written the critical sense of physicists had not been sufficiently matured to permit a scepticism as extreme as this, and statistical mechanics was developed in the conviction that this extrapolation *must* have meaning and be legitimate.

The first results of the statistical approach were indeed most heartening. The conservation of energy of an isolated system became at once the known theorem of the conservation of energy of a purely mechanical system; the interconvertibility of thermal and mechanical energy were at once understandable in terms of Rumford's picture of heat as consisting of the mechanical motion of submicroscopic particles, and the universal tendency of mechanical energy to dissipate itself into thermal energy was understandable as an illustration of the tendency of the energy of a mechanical system to ultimately distribute itself uniformly among all the degrees of freedom. This latter was a doubtful triumph, however, because when applied to electromagnetic systems it said too much, and led to the dilemma of the "ultra-violet catastrophe," which historically was the occasion of the introduction of the quantum by Planck.

Physicists probably first became generally conscious of some of the difficulties in the statistical picture through the invention by Maxwell of his "demon." Maxwell pointed out that an intelligent being small enough to be capable of dealing separately with the motion of individual molecules would be capable of violating the second law. A simple method of doing this in a system consisting of a gas at uniform temperature in an isolated enclosure would be to put a partition down the middle of the enclosure provided with a trap door, which would be opened and closed by the demon to permit the passage of molecules from left to right whose velocity was greater than average and from right to left whose velocity was less than average. In this way the right hand half of the gas would become continually hotter and the left hand colder without the expenditure of any energy. After the temperature difference had been built up, it could be used to drive a thermal engine, which would deliver useful work, at the ultimate expense of the energy of a body of gas at uniform temperature, in violation of the second law. This argument was at the time accepted at its face value as indicating the need for some drastically new addition to the kinetic picture before it could be regarded as affording a satisfactory explanation of the mode of operation of thermodynamic systems. One of the ways out, and that usually accepted as the sound one, was through the introduction of various aspects of the concepts of probability and randomness. An examination of the implications of this will concern us presently. But there were other consequences of the recognition of the possibility of the existence of demons which left their impress on physical thinking. If this possibility is seriously entertained, it throws the whole situation in thermodynamics wide open as far as practical consequences are concerned. For what does the practical man care by what method he produces perpetual motion of the "second kind" so long as he gets it? He would be perfectly willing to use as the instrument for getting it a culture of bacteria or an army of demons shut up in a box, provided with the proper marching orders.

If the Maxwell demon had been invented yesterday instead of in the last century I believe he would not have caused as much consternation. There are too many vital points that must be cleared up. In the first place, what is this "intelligence" that must be presupposed? Even if it need be nothing more than the ability to make automatically a selective response to the velocity and direction of the approaching molecules, such an ability involves some sort of mechanism, and this mechanism must have a minimum size and number of molecules and might well be so coarse as to make dealing with individual molecules impossible. I think we would not now be willing to assume such a mechanism unless we could give some suggestion as to how to construct it. But Maxwell apparently was willing to endow his demon with a gray matter composed of material of fundamentally different properties from any known; it almost looks as though he took a "mind stuff" seriously. I think we now would not be willing to assume this possibility; we would be afraid that in some way the properties of matter as we know it are connected with the non-existence of such undiscovered kinds of matter. Another doubtful feature is the method by which the demon would learn of the approach of the individual molecules. The only method would appear to be by light signals; these must come in quanta and must react with the molecule. The reaction is uncontrollable, and may be sufficiently large to divert the molecule by so much as to vitiate the manipulations of the trap door. Then there is the question of the legitimacy of assuming that "information" can be propagated in any purely thermodynamic system. The propagation of information demands a differentiated radiation field, whereas in a body at uniform temperature the radiation field must also be uniform. Again, there is the question of the details of operation of the trap door; if the door is a mechanical system in the proper sense it must be perfectly elastic, and this means that its motion cannot be stopped, or is at most periodic. In other words, how can a mechanical system be set into motion and braked again at arbitrary intervals without dissipation? Closely connected with this question of the operation of the trap door is the question of thermal fluctuations: when the mechanism gets small, the mechanism itself and its controls (including in the controls the brain of the demon) become subject to temperature fluctuations which are proportionally larger the smaller the mechanism. How do we know that this will not vitiate the entire program?

These are serious questions, and so far as I know have received no adequate discussion. These difficulties, taken in conjunction with the fact that the entire invention of the demon is most obviously a paper and pencil affair, would, I think, limit the influence of the invention of the demon to the paper and pencil domain if it were not for a couple of facts of the laboratory. In the first place, it is possible actually to see with our own eyes the fluctuation phenomena which are the basis of the operations by the demon, as anyone who has seen the Brownian motion in a microscope or projected on a screen knows. In the second place, it is possible to bring up these small-scale things to the control of the events of ordinary life; by the use of a Geiger counter and amplifier a battleship may be destroyed by a single radioactive disintegration. The result of all this is a genuine doubt in the minds of many physicists as to whether it will sometime be possible, by the construction of sufficiently clever apparatus, to violate the second law on a commercially profitable scale. This doubt is probably considerably modified in the minds of those who have seriously assessed the significance of the fundamental postulates of quantum theory.

A proper understanding of the situation demands that we understand the essential difference between trying to get energy out of fluctuation phenomena and out of the waves of the sea. We can accomplish the latter by the use of ratchet mechanisms; why do we assume that ratchet mechanisms are impossible on the scale of thermodynamic fluctuations? We have already emphasized that the assumptions of statistical mechanics are nothing but extrapolations from everyday experience; by what right then do we assume that the ratchet mechanisms cannot be extrapolated?

It seems to me that the extrapolation we make in statistical mechanics is really not an extrapolation of everyday experience, but of the mathematics of everyday experience. The Lagrangian equations of mechanics cannot handle ratchets; we extend the Lagrangian equations to the world of particles and should not be surprised that we find no ratchets awaiting us.

Granting now that the sort of microscopic control needed to make the demon effective is at present beyond our attainment, let us examine the argument by which the kinetic picture is adapted to the actual world about us. The method of course is to add the notions of probability, converting a straightforward mechanical situation into a statistical one. The whole question of the logical analysis of the concepts of probability is one of the most difficult there is, and there is still much controversy even on fundamental matters. But there are certain aspects that are straightforward enough. There is a clean-cut mathematical discipline of probability; there are rules by which the probability of compound events may be calculated from the probability of simple events, and once given the formulation of any problem, all mathematicians would agree on the numerical result. That is, there is a clean-cut paper and pencil subject of probability, with clean-cut operations. The difficulties and disagreements arise in making practical applications, and in connection with questions like the following: Do the fundamental operations of the paper and pencil subject sufficiently correspond to what occurs in practical situations? Can we get right answers in practical cases by our paper and pencil operations? If we can get the right answers, what is the proper method of preparing the actual situation for paper and pencil attack? If we can get the right answers, what is the reason that we can get the right answers? This last comes pretty close to a meaningless question; in situations where we feel that we have given a "reason" I think that analysis will show that we have only been able to exhibit a parallelism between a "physical" situation and a succession of paper and pencil operations which we feel to be particularly close. Our question then becomes this: in those cases where experience shows that we can get the right answer by the mathematical operations of statistical theory, how close is the parallelism between the details of the operations and of the physical situation?

In spite of disagreement with regard to many fundamental questions of application of probability notions to actual situations, I think there is agreement on certain points. For example, I think it is coming to be more and more accepted that the notion of a probability is not applicable to any single concrete individual situation. Before an individual event happens, the only meaning that can be attached to its probability concerns a state of mind with regard to it: either a program of action on my part or a schedule of previous events similarly embedded in their surroundings. When the event occurs all is altered; we have one definite thing, and that is that: either the die comes a six or it comes a three, etc. This is all recognized in the popular role that the ensemble of systems has come to play in recent treatments of probability. It seems to me, however, that the logical difficulty is thereby only shoved into the background; an ensemble of systems is itself only a larger single individual thing. The fuzziness which makes a system, either an individual or an ensemble, amenable to statistical treatment does not reside in the system as such, but is superposed upon the system by fiat. A system is a probability system if I treat it by probability methods; this makes it necessarily a paper and pencil affair.

THE ENSEMBLE AND EPISTEMOLOGY

The use of the ensemble has particularly become fundamental in wave mechanics treatments of probability. In fact a virtue has been made of this use of the ensemble, which has been incorporated into a system of epistemology. According to this epistemology, the ensemble is the only proper thing to use, because we never have physical knowledge of an individual, but only of an ensemble. There is of course much truth in this point of view. The equations of physics purport to deal with physical "laws," that is, with general as distinguished from particular situations. The v of the perfect gas law is the v of *any* volume of gas, and the numerical value of v which is to be substituted into the gas equation is the v obtained by measurement of any body of gas. Furthermore, according to this point of view the v which may appear to be obtained by an individual act of measurement on a specified body of gas is more properly to be regarded as obtained by a whole ensemble of operations. For no single measurement is ever accepted in a physical laboratory as a proper measure, but must be checked by repetition. This is both to guard against possible "error" in a single measurement and to ensure greater accuracy, for every physical measuring instrument has divisions of only limited fineness, and the fraction of the finest division is something which has to be obtained by an estimation, which is not likely always to repeat itself (this is where the notion of an ensemble fits with particular nicety), and in the result of which one feels greater confidence the greater the number of individual readings that make up the average. This aspect of our ordinary laboratory measurements is similar to an aspect of the ensemble.

Further, the physical experience which is epitomized in the mathematically formulated law was itself not a single experience but an ensemble of experiences; the law could not be formulated until we had had experience of a wide range of individual situations. That is, every law is based on extensive previous rehearsals, which we have already seen are necessary in evaluating the energy and entropy functions of thermodynamics.

There is still another respect in which use of the ensemble corresponds to the character of knowledge. From a study of the statistical properties of an ensemble one never arrives at an absolute certainty. Similarly, one could take the position that one can never say that any event of daily life, even such a universal event as the falling of a stone to earth when its support is removed, will *certainly* occur. All one can say is that there is a very high probability of such an event, but the contrary event is "possible," and one might even undertake to give numerical limits for its degree of probability. Such a theory of knowledge demands a considerable sophistication, however, because in practice we unquestionably deal with the falling of a stone as if it had complete certainty. That one can nevertheless make such a theory go appears when one considers the numerical values that would have to be given for the probabilities of such events as the falling of a stone upward against gravity. These probabilities are so fantastically small that even in all the history of the human race the "chances" are very small that such a thing has happened, and of course they are still smaller that any individual will observe such a thing in his own lifetime. When applied to the description of known events in the past or to my program of action for the future such a theory will therefore be indistinguishable from the more congenial sort of theory that appears to ascribe absolute certainty to the fall of the stone downward. Now as a matter of fact, such a certainty, absolute in *all* its aspects, is never ascribed to any prediction of ours about the future, for we would always be willing to *say* of any prediction that it is not absolutely certain, and we have to recognize that the stone "*might*" fall upward. So that there is this *qualitative* correspondence between our common-sense actions and a theory of knowledge which assigns a small numerical probability to events so "rare" that they have never been observed. A theory which is used so as to exploit this correspondence may therefore be useful. But I think that the correspondence is not so far-reaching that one can completely substitute the one for the other. As long as we are concerned only with the calculation of numerical values and their use in laying out programs of action in practical cases there may be equivalence between the two sorts of epistemology. One might be tempted to say that "operationally," since there is nothing to distinguish between the two views, there is in fact no difference, and that they must be regarded as the same. This would indeed be true, only "operationally" must cover *all* the operations, not merely those of numerical calculation and program drawing. In particular, the operations include the entire verbal structure into which the two views may be fitted. From this point of view there is a difference which must not be overlooked between the common-sense point of view which says that a pail of water has never frozen on the fire and I shall make my programs for the future without leaving any room for this actually happening although I recognize it as a *possibility* that it *might* happen, and the point of view that says that there is a chance of 10^{-1000} that a pail of water might freeze on the fire but this chance is so small that it would be waste of time to figure what course of action I had better adopt if it did happen. The difference between these two points of view appears when one analyzes what one means by saying in the language of every day that it is *possible* that the pail *might* freeze on the fire. This verbalization I think is merely a way of giving recognition to that one characteristic of our method of dealing with the future which we would maintain to the bitter end. This is that "there is no appeal from experience." It is what happens to us that must control, and not our paper and pencil schemes or our most cogently argued expectations. If we find stones falling upwards, that is that, and all there is to it. This I believe is all that our everyday epistemology can mean when it says that it is *possible* that the pail *might* freeze on the fire. The verbalism itself and the implications for *verbal* behavior in other situations are entirely different from saying that there is a 10^{-1000} chance that the pail will freeze. It seems to me therefore rather accidental that the statistical epistemology gives room to a *qualitative* correspondence with the fact that the epistemology of common sense also says that predictions of the future never have absolute certainty. One is not for that reason to jump to the conclusion that an epistemology based on statistics is more "inherently correct" than the epistemology of common sense in its more naïve forms which talks about absolute certainties in certain situations.

The same sort of argument can be applied to the contention that all our knowledge is at bottom ensemble knowledge; this may be justified as far as many important aspects of our operations are concerned, particularly the aspects dealing with arithmetical computation, but it is not justified when we extend the operations to include the complete verbal structure. For we do *say* that our past experience consists of individual events each of which has its own identity; we must think of even the individual members of any ensemble as having identity. We shall talk about and think of our future as consisting of individual events, each with identity, none of which ever occurred before and none of which will ever occur again, and we demand of our physics that it enable us to deal with each of these individual events as it occurs. It is true that our physical laws are formulated in general terms applicable to the members of an ensemble, but this is only because of convenience, because by this sort of a formulation we can deal with any individual whatever, no matter what it is. Our problem and our concern remains with the individual thing and the individual event.

All of this does not question at all the great utility for physics of a statistical epistemology. Such an epistemology lends itself to numerical computation much better than the epistemology of common sense, and since numerical computation is such a large part of the business of physics, it may well be that the physicist will adopt this and discard the other. If he does, we may expect that presently he will be urging that the verbal impulses back of the epistemology of common sense are not trustworthy impulses, and that they are to be resisted rather than followed. It is clear enough that the possibility that such a contention is justified must at least be recognized. I am not at all clear as to the eventual outcome; it will have to be settled by experimenting to determine whether such verbalisms as "The events of the past have an identity which they retain forever" are as indispensable as we had thought, or whether substitutes can be found for them. It is even not inconceivable that the one epistemology will be retained for daily life and the other for physics. It is not easy to see how one would construct a mathematics corresponding to the "whole or nothing" epistemology of daily life. If two epistemologies are retained, there will be a certain justification for talking about an essential difference between the point of view of science and that of common sense. But in any event, one will, I think, be unwilling to talk of this or that epistemology as being "the" "correct" one, a designation which some physicists are showing a tendency to believe can be applied to the statistical epistemology. I believe that no epistemology can be logically rigorous, but between rival epistemologies it can only be a question of which is logically most tolerable in a particular setting. For the meanings of the words in which any epistemology is framed demand a break-down of situations into parts isolated from each other and which exactly recur, and this break-down is only approximate.

ENTROPY AS DISORDER

Turning now to more specific questions, perhaps the most important result of the statistical-kinetic picture is the account it gives of entropy as determined by the amount of "disorder" in a system. It is not easy to give a logically satisfying definition of what one would like to cover by "disorder," and it has been a favorite topic for discussion. The naïve idea seems simple enough. Everyone knows what it means to "shuffle" a pack of cards, and would be willing to claim that the cards for a particular hand had or had not been "well shuffled." But the situation is not so plain when one begins to look at it. Can one describe the shuffling as "good" when only the *operation* of shuffling is specified, or must one also know the result of the shuffling before one can say that it was good? One natural definition of a "good" shuffling is one that puts the cards into disorder, and whether the cards are in disorder or not can be told only by an inspection of their actual distribution. But G. N. Lewis has justly pointed out that it would be possible to formulate the rules of some card game so that any arrangement of the cards whatever would be a regular arrangement from the point of view of that game. "Disorder" is therefore not an absolute, but has meaning only in a context. What is the context which gives meaning to the disorder of the physicist when he talks about entropy as a measure of disorder?

The context, it seems to me, can be nothing less than the whole complex of operations which we employ in thermodynamics. These operations may be operations of the laboratory or pencil and paper operations. In the first case the operations are performed by us or our fellows; in the second, the formulation of the operations must be made in human language. The position to which we are thus brought has been very distasteful to some because of the "anthropomorphic" element. It is felt that a concept so universal and fundamental as entropy should not involve elements so patently anthropomorphic. This consideration appealed very strongly to Planck, for example, whose discussion of entropy was inspired by the desire to free it from its anthropomorphic elements. This it seems to me cannot possibly be done, using "anthropomorphic" in Planck's sense. Thermodynamics itself, I believe, must presuppose and can have meaning only in the context of a specified "universe of operations," and any of the special concepts of thermodynamics, such as entropy, must also presuppose the same universe of operations.

It seems to me that the necessity for a universe of operations is shown with special clearness by an analysis of the Gibbs paradox presented by the diffusion of a gas into itself. The paradox is well known. Given a box, with a partition in the center, and on the two sides of the partition two different gases: The partition is removed and the two gases allowed to diffuse into each other. Complete mixing is accompanied by a characteristic and easily calculable increase of entropy, the same no matter what the nature of the two gases. Hence the increase of entropy must remain the same in the limit when the two gases become identical with each other. Here is the paradox, because the parts of a single uniform gas are continually inter-diffusing into each other, and there is no increase of entropy.

To resolve the paradox, we must recognize in the first place that we have here a question of double limits and that the limiting process has been improperly performed. In the original set-up the universe of operations includes the operation by which the gases originally in the two halves of the box can be distinguished from each other, i.e. the operation by which two "different" gases can be distinguished. It is assumed that this operation can still be performed and has meaning during all stages of the limiting process, as the two gases are allowed to become more similar. Of course this condition becomes more and more difficult to realize instrumentally, but verbally the formulation remains the same. But in the final step of the limiting process there is an abrupt discontinuity in the instrumental process; the two gases now become "identical," which means that there is no instrumental operation performable *now* by which they may be distinguished, but the instrumental operations by which two molecules are to be distinguished would demand a determination of the history of each molecule, according as it was originally in the right or left hand side of the box. Such operations are only pencil and paper operations, not the laboratory operations on which thermodynamics is predicated and which give entropy its meaning. Since the universe of operations of the limit differs discontinuously from that of the preceding stages, the argument for the continuity of any function, such as the entropy, falls.

The entropy in general therefore must depend on the universe of operations and must change when the universe changes. One is tempted to paraphrase Patrick Henry and say "If this be anthropomorphism, make the most of it."

There is a suggestive parallelism between the abrupt change in the assignment of entropy which follows a change of operations which from one point of view might be described as an infinitesimal change and the abrupt change in the assignment of localization to energy which we have seen may follow the discovery of a new kind of instrumental operation, no matter how apparently insignificant.

We have seen that the concept of "disorder" cannot help being "anthropomorphic," and we have particularized this by showing that entropy can have its meaning only in a setting of a universe of operations, but we have not considered the relation of the disorder to the universe of operations which gives entropy its macroscopic thermodynamic meaning. It is pretty evident, I think, that the disorder is not to be found in the realm of the macroscopic operations which give entropy its thermodynamic meaning; these are operations with the instruments of the laboratory, pressure gauges and thermometers, and mathematical operations with pleasant analytical functions. It is hard to see anything of disorder here. The disorder is obviously in the paper and pencil domain, and therefore palpably and necessarily "anthropomorphic." But what is the meaning of a paper and pencil disorder? To answer this question would demand an analysis very similar in most respects to the analysis of the meaning of the concept of probability. I have already considered this latter with some care, and shall not repeat the discussion here. Perhaps the most important conclusion of that discussion was that there is no logical method of bridging the chasm between the verbal, paper and pencil structure and our application to concrete cases. It is not that a similar chasm is not always present whenever we attempt to justify the application of any theoretical argument, whether involving probability or not, to a concrete situation; it is only that the chasm is a little more yawning and obvious in the situations presented by probability than in more classical situations. The application of ideas of probability or of "disorder" can be made to concrete situations only by a logical quantum jump. The procedure of making the jump would be neither useful nor justifiable were it not a procedure that we can perform unambiguously and easily. That we can proceed thus unambiguously and easily is a trait of the human animal the existence of which is a matter of simple observation. All persons who have considered the matter would go to work in the same way to find how to bet in a given situation, nor would there be any hesitation in saying that a large number of white and black marbles which had been tumbled about in a barrel for a considerable time would present a uniformly gray aspect from a distance. This unambiguity of procedure is all that is necessary, not its logical justification.

Simple observation shows that there is this unambiguity in attaching paper and pencil meaning to the concept of disorder in the situations presented by statistical kinetic theory. We have no hesitation in saying that the molecules of a gas which were originally in the right and left halves of a box will presently be uniformly mixed, or that pressure differences will smooth themselves out, or temperature differences. These qualitative systems are handled confidently and unambiguously.

Although not necessary for our purpose, I think it is interesting to inquire if we cannot give some account of this ability of ours. It seems to me that perhaps the most important factor here is a variant of the principle of "sufficient reason." If the right and left hand portions of gas do not eventually mix uniformly, there must be some parts of the box that have a preponderating proportion of one or the other component. There must be some "reason" for this, but the kinetic picture indicates nothing in the original distribution which could serve as sufficient reason for the final preponderance of either component in any part. The original positions and velocities were so distributed that there is no reason why any recollection of them that can be specified in terms of macroscopic operations should persist indefinitely. "No reason why" is equivalent to saying "disorder."

We would not be able to handle this situation so unambiguously if we did not have so many practical examples in which the amount of disorder in an assemblage of many objects is observed to increase with the passage of time if it is exposed to external stimulation or agitation, as in our barrel of marbles. This is the obvious and striking analogue of the tendency of entropy to increase, and provides the natural motivation for the identification of entropy with disorder. It is strange that we do not seem to require any explanation for the tendency of a system of many members to increase in the disorder of its arrangement, but this tendency is such a universal property of the systems of ordinary experience that we know intuitively when to expect it and do not require any explanation, unless we are unusually critically minded. If we do hunt for some justification for our expectation it might perhaps run as follows. Suppose, for instance, we open and close a stopcock allowing a jet of gas to discharge into an evacuated chamber. We are morally certain that presently the molecules of the gas will be uniformly distributed throughout the enclosure, bouncing back and forth with a completely "haphazard" distribution of velocities, in spite of the fact that initially the positions and velocities were far from random. We find the mechanism for the diffusion of position and velocities in the collisions of the molecules with the walls, the irregularities of which have no significant correlation with anything in the initial structure of the entering jet. If we are asked how we know that there is no "significant" correlation, our reply is almost certain to be that we can see no reason why there should be such a correlation. We fortify our position by analysis of the details of manufacture of the chamber and of the known properties under other conditions of the material of the walls and of the gas. Anyone would certainly have to admit that as we tell our story no hint of correlation appears. Our position thus may be made to be entirely satisfactory as far as naturalness and workability go, but this cannot conceal its logical weakness.

It would thus appear that disorder in the paper and pencil sense and on the atomic scale implies no *significant* correlation between the array which we are considering and other arrays with which it may interact, thus modifying its behavior. In any event the disorder must be of one thing with respect to another; as an absolute term, disorder has no meaning even in the paper and pencil domain. And when we say "significant" correlation, we imply a context which imparts the significance. For the thermodynamic applications the context which gives the significance is the large-scale macroscopic operations of thermodynamics. In the example of our jet above, there was no correlation between the details of the original molecular distribution in the jet and the irregularities of the wall which was significant for any measurements that could be made with pressure gauges and thermometers, and whose results could be formulated in terms of measurements made with such instruments.

It is not possible to say that there is no correlation between the small-scale irregularities of the wall and large-scale indications with pressure gauges and thermometers without a theory of some sort. Part of the context which gives disorder its meaning therefore involves a theory. The precise nature of the theory does not actually play a very important part. In the case of the gas molecules impinging on the solid walls of the container, the theory assumes the ability to write the equations of mechanics for every detailed encounter, but makes no pretense of carrying through the exact solution. Instead, the necessity for a detailed solution is waved aside with a large gesture and a reference to the tremendous complexity of the boundary conditions which make any large-scale correlation "infinitely improbable" — again the argument of sufficient reason.

The large-scale operations with which the correlation is to be found that give "disorder" its technical meaning are all of a special kind. These are obviously all operations by which energy may be extracted from a system, for the second law is formulated in terms of the work which may be extracted. Thermodynamics has nothing to say about other types of order or disorder which in their own context may be of the greatest significance. For instance, thermodynamics makes no attempt to give a rating to the relative order of the paintings of a da Vinci or a Turner, and certainly does not attempt to attach an entropy to a work of art. In fact there is a school of art, that of the Futurists, which would use "disorder" in a sense to give exactly the opposite results from thermodynamics. It is said by some that this school sees the prototype of beauty in anything of a complete naturalness. Hence to this school the drive of the universe toward the complete chaos of molecular disorder must appear as the height of all beauty, and therefore of the highest order.

Even in the domain of situations which would be unanimously described as primarily "thermodynamic" it would appear that there is a rather large verbal element in the coupling of "disorder" with entropy, and that this coupling is not always felicitous. Consider, for example, a quantity of sub-cooled liquid, which presently solidifies irreversibly, with increase of entropy and temperature, into a crystal with perhaps a regular external crystal form and certainly a regular internal arrangement as disclosed by X-rays. Statistically, of course, the extra "disorder" associated with the higher temperature of the crystal more than compensates for the effect of the regularity of the crystal lattice. But I think, nevertheless, we do not feel altogether comfortable at being forced to say that the crystal is the seat of greater disorder than the parent liquid.

There are difficulties at the absolute zero if we attempt too close a correlation of the common-sense notion of disorder with entropy. According to the third law of thermodynamics, entropy goes to zero at zero degrees, and therefore there can be no disorder left. How are we to calculate the disorder to associate with the large-scale geometry of a system? In a room at absolute zero are the shapes of the modernistic furniture to be described as perfectly orderly, and if not, are we to anticipate that if we wait long enough these shapes will perhaps presently smooth themselves out spontaneously into something uniform and amorphous? One would rather look for such a spontaneous transformation at high temperatures. The answer must be that the "disorder" which is the occasion of entropy is not a disorder in the parameters in the "mechanical" group. It is rather a disorder in the group of "sub-mechanical" parameters which are the origin of the "thermal" behavior and properties of a body. From the point of view of thermodynamics such parameters are entirely in the paper and pencil domain. This is as it should be, because "disorder" is not a thermodynamic concept at all, but is a concept of the kinetic-statistical domain. A problem which the kinetic-statistical point of view has to solve is where to draw the line between "mechanical" and "sub-mechanical" parameters. As always, I do not believe the line can be made sharp. This again is to be expected. There is a fuzziness about the common-sense notion of "disorder" which makes it not always altogether suited as an intuitive tool in discussing the second law.

It is natural to attempt to extend the notion of disorder to give a method for distinguishing between mechanical energy and energy in the form of heat. The sentries which we have had to post so many times at the boundaries of a region to which we are applying the first law would now take the form of Maxwell demons, and we would instruct them to report all "disordered" energy which passed them as heat and all ordered energy as "mechanical." The sum of the fluxes of "ordered" and "disordered" energy must add to the total flux of generalized energy which we have seen has an independent meaning in terms of macroscopic operations. The picture thus presented has perhaps some slight value, but it presently leads to formidable difficulties. For instance, imagine a current of heat flowing by conduction down a temperature gradient. There is an element of "order" in this flux of heat due to its unidirectional character. How much of the order which the demons observe in a case like this shall we instruct them to classify as "heat"? Similarly, temperature might be defined in terms of the "disordered" energy, but the definition is largely verbal and almost entirely in the paper and pencil domain. It is not obvious how with this picture of heat and temperature one would deal with an evolution of heat at absolute zero, for the "heat" of irreversible transitions does not vanish at zero.

So much for the qualitative ideas that entropy corresponds to disorder, and that the natural tendency for entropy to increase corresponds to the natural tendency for disorder to increase. Incidentally this involves the thesis that it is not possible to construct for atomic phenomena sorting machines analogous to those which can be constructed for large-scale things. There are difficulties in seeing why this is impossible which bring up all the difficulties presented by the Maxwell demon. I find that these difficulties dim my intuitive ease in handling the picture in new situations. Accepting, however, the qualitative picture, the next step is to make it quantitative by finding some way of measuring numerically the amount of disorder and thus getting a precise parallel with entropy or even a method of calculating entropy. This will necessarily involve some modification of the naïve notion of disorder. A complete discussion of the methods used here would take us too far into the technicalities of statistical mechanics. The fundamental and sufficient point for us is that the problem is reduced to one of counting the possible arrangements of the system. The general idea is to compare the number of arrangements corresponding to any one macroscopic condition with that corresponding to some other, and thus get a comparison of the entropies of the two conditions. The difficulty is that counting processes cannot be applied in systems which are capable of infinite subdivision; counting requires that the process of subdivision stops somewhere. It turns out that it does not make much difference where the subdivision stops, provided that the size of the elements is sufficiently beyond reach with macroscopic instruments, because the ratio of the counts for any two specific states of the system will be approximately independent of the size of the elements. Accepting any arbitrary size for the elements of "phase space," it then is proved in the classical statistical mechanics that it is possible to get a quantitative expression for entropy in terms of the amount of "disorder." From the paper and pencil point of view the problem may therefore be regarded as solved because we have a formal method of calculating the entropy in any specific instance. But from the "physical" point of view the situation is very unsatisfactory, because we would like to have some physical indication of what the size of the elements of volume should be, and in the realm of classical phenomena there is no indication that there is such a size. We like to feel that our picture of entropy as a measure of disorder is based on some deep-seated physics, but as we work it out there is nothing but a paper and pencil structure, with a very fortunate parallelism with the physical situation, which we exploit.

With the advent of quantum theory and the discovery of quantum phenomena the situation is fundamentally altered. Physical phenomena now do set a natural size for the element of volume in phase space, and definite numbers may be assigned to the number of physically significant arrangements corresponding to some macroscopic state. The temptation is to project our present knowledge into the past and to say that the old intuition was essentially correct, and that because it has turned out all right, our previous failure to find a natural element of volume was a matter of no moment. It does not seem to me, however, that we have any right to such complacency. Quantum theory puts the notions of probability in at the very beginning, instead of trying to derive them at a later stage as a consequence of something more fundamental. The result, it seems to me, is that quantum statistical theory is fundamentally an entirely different sort of structure from classical statistical theory, in spite of the close parallelism and formal mathematical similarities. The proof can be found by an examination of the operational meaning of the probabilities used in the two theories. It follows that a questionable procedure in classical theory cannot be justified by finding in quantum theory a parallel for it which is capable of more or less rigorous justification.

CHAPTER III

MISCELLANEOUS CONSIDERATIONS

THE UNIVERSE OF OPERATIONS OF THERMODYNAMICS

WE HAVE seen that thermodynamics presupposes a "universe of operations"; what is this "universe of operations"? One has to distinguish as best he can between instrumental operations of the laboratory and paper and pencil operations. It is obvious enough that not all possible paper and pencil operations are allowed to thermodynamics, for otherwise there would be no basis for the distinction between the thermodynamic and the statistical-kinetic points of view, which we may suppose is real because it at least is now affording the subject for our discussion. But with regard to laboratory operations, one may ask why there has to be a universe of operations; why should there be any restrictions, or why should not our universe of laboratory operations be "all the operations that we can perform"? Such a catholicity is certainly permissible, with the proviso that we should have said "all the operations which we can *now* perform"; the only question is whether we would have wanted to call what we had got thermodynamics. It is evident enough, however, that as a matter of fact there are limitations on the laboratory operations which we permit in thermodynamics. Perhaps most important, nearly all thermodynamics assumes the permanency of the atoms of the chemical substances with which it deals. This is in spite of the fact that we are now finding how to change one atom into another and are coming to believe that probably any specifiable permutation is possible under conditions which perhaps we may learn to control. We originally did not recognize this possibility and built a thermodynamics which was consistent and profitable, in spite of the fact that its universe of operations was one from which the operation of transmutation was eliminated. There is no reason why another discipline, which we may call a generalized or extended thermodynamics if we like, should not be built in which the universe of operations does include transmutation, and in fact beginnings of this more general subject have already been made. The reason that it was possible to build a consistent thermodynamics without transmutation was of course that the act of transmutation was one which occurs spontaneously so infrequently that we were not ordinarily aware of it, so that as long as we could not make it occur of our own volition, it never obtruded itself.

There are at least two other restrictions on the operations permitted to classical thermodynamics in addition to that on transmutation: these are restrictions on the scale of the instruments and on the time necessary to obtain the readings. There must be at least a paper and pencil restriction on the scale of the instruments, for otherwise we would be led straight into statistics and kinetic theory. It is perhaps not so obvious that there is a laboratory restriction on the scale of the instruments. Indeed, in the early stages of thermodynamics, when Kelvin spent years establishing the existence of the Thomson effect with mercury-in-glass thermometers, it would not have occurred to one that it might be necessary to place a formal limit on the size of his instruments, but every advance in the direction of the small would have appeared to be an advantage. Kinetic theory, however, with its dictum that the "temperature" of an individual molecule meant nothing, indicated a theoretical limit to the smallness of the permissible instruments. This limit was presently caught sight of by the skill of instrument makers, first in the optical demonstration of the Brownian motion, then mechanically in the spasmodic motion of light objects such as galvanometer suspensions, and then electrically in the "small shot" effect in vacuum tubes. The experimental demonstration of these fluctuation effects exerted a powerful stimulus on the imagination of physicists and hastened the general acceptance of the kinetic picture. I believe that the enthusiasm evoked by the revelation of the world of fluctuation phenomena to a certain extent over-reached itself and resulted in an uncritical acceptance, lock, stock, and barrel, of all the implications, both instrumental and verbal, of the kinetic picture.

At any rate, it is evident that the instruments of thermodynamics must be of a scale so large that they do not respond to these fluctuation phenomena, for otherwise systems would never come to equilibrium, which is the fundamental assumption of the temperature concept and of all thermodynamics. But this lands us in a difficulty, because there is no sharp limit which separates the size of an instrument sensitive to fluctuations from one irresponsive, but the more sensitive we make our methods of reading our instruments the larger the instruments we find exhibiting fluctuations, and our paper and pencil operations assume that even objects as heavy as the instruments of the kitchen or garage have fluctuations whose amount can be specified numerically. The situation is not, however, for this reason any different from what we find in all the rest of physics.

We never have sharp dividing lines, but we do have verbal rules by which we sharpen the fuzzy results given by our instruments. In the case of thermodynamics we sharpen our instrumental data by the requirement that the numbers we substitute from the readings into our calculations shall not exhibit fluctuations, but shall vary smoothly both in space and time. "Fudging" the data in this way may seem a scandalous thing to do; it is nevertheless justified and is at the bottom of all scientific procedure. The justification of course is that in this way we acquire command of regularities in the behavior of the systems about us which are of importance. We cannot proceed arbitrarily, but there are definite conditions that have to be met. What are the conditions that we impose on the "universe of operations" of any scientific discipline, in particular thermodynamics?

One condition that we always try to impose when we can is that the system which we generate with any particular universe of operations must be a "causal" system. That is, when a system is completely described in terms of all the operations in the given universe, then the future course of the system, described in terms of the *same* universe of operations, must be completely determined, or, in other words, it must repeat itself when the prior conditions are repeated. Applied to thermodynamics, this merely means that the thermodynamic parameters, determined by instruments and operations within the thermodynamic universe of operations, must be adequate to permit the existence of the "science" of thermodynamics. It is a matter of experiment that it is possible to delimit a universe of operations of this kind; the existence of the science of thermodynamics is proof of the possibility. Here, as always, we have nothing sharp, but are able to satisfy the conditions only approximately. The conventional operations of thermodynamics do not set apart a system of phenomena which are completely causal in the sense of our definition, because even in our ordinary world in the reach of our large-scale instruments and operations, events sometimes occur for which there is no causal explanation inside the system, as for example the blowing up of a battleship by a single atomic disintegration, coupled to a Geiger counter and an amplifier. It is only that such events are so rare that we can profitably disregard them for a great many of the purposes of ordinary life. Neither is it possible to set any sharp numerical criterion which determines at what point the approximation is to be considered unsatisfactory and an enlargement sought in the universe of operations which shall be more satisfactory.

The instruments of thermodynamics include thermometers and instruments for determining the various mechanical parameters, such as pressures or stresses or electrical or magnetic fields. They must not be large compared with the geometry of the boundaries of the systems we have to deal with, and they must be small enough so that with the help of them the given system can be analyzed into elements each of which is sensibly homogeneous. The meaning of "sensibly homogeneous" has a large paper and pencil component: for instance it is required that the sum of the energies of the elements regarded as homogeneous be equal to the total energy, and similarly for the entropy. Within these restrictions a great deal of latitude is possible in the actual scale. The situation is the same as that of which Lorentz early made physicists conscious with his discussion of "mathematical" and "physical" infinitesimals. As the size of the instruments is diminished, the data first pass through wide fluctuations imposed by the gross geometry of our system; that is, at first a single instrument may be trying to straddle a piece of iron and a piece of copper. As the instruments get smaller their indications smooth out and approach a smooth level plateau. As they get still smaller, fluctuations again begin to manifest themselves. The universe of thermodynamic operations is restricted to the region of the plateau. It is a matter of experiment that there is such a plateau. It is also a matter of experiment that the theater of such a universe of operations is a sufficiently close approach to a causal system to be profitable.

The sort of "causality" with which we are satisfied in thermodynamics is less complete than a logically rigorous causality; such an incompleteness is a natural corollary of our recognition that the boundary line is hazy between the macroscopic operations of thermodynamics and microscopic operations now possible in the laboratory. Thermodynamics is concerned with quasi-equilibrium states sufficiently characterized by macroscopic operations, separated from each other in general by other states which we know cannot be adequately characterized by such operations. These intermediate states constitute by far the major part of our experience. We are satisfied to call a system "causal" thermodynamically if the succession of quasi-equilibrium states is repeatable. For example, an imperfect gas expanding through a nozzle into a final volume twice the initial volume passes through intermediate states of hopeless complexity, but the initial state is adequately described by a few simple parameters as is also the final state. It is an experimental fact that the final state is repeatable and predictable, given the initial state, so that the final and initial states are "causally" connected. Note incidentally that among the parameters necessary to make this a causal system are the geometrical parameters describing the shape of the chambers and the orifice through which expansion takes place, obviously macroscopic parameters.

The instruments of our universe must include others than those necessary to determine the numerical values of the various thermodynamic parameters. There must be instruments for giving qualitative descriptions of the objects of the system. If we have no way of knowing whether it is a piece of iron or of copper which is being inserted into a solenoid, our system will hardly appear to be a causal one. Evidently there is a great deal of latitude here. If, for example, the identity of the various pieces of matter in the system can be established, then a set of operations which describes the history of the behavior of any of the individual pieces of matter over a sufficiently wide and varied past would be capable of replacing operations of a chemical character which determines the nature of the pieces of matter by measurements performed now.

Although the operations of thermodynamics may include a determination of various qualities of an object, complete characterization of *all* the perceivable qualities, such for example as artistic merit, is not contemplated, but only those qualities which are pertinent to the ultimate purpose of thermodynamics. The broadest characterization of this ultimate purpose is perhaps that it is concerned with all possible manifestations of "energy," a verbalism which is bad enough, but which nevertheless suggests the situation perhaps sufficiently well. From this point of view the determination that the color of a particular object is yellow would not ordinarily be described as an operation of thermodynamics, but it may become one if the object for example is mercuric iodide, and the yellow color is associated with the fact that at 127° the color changes from red to yellow with absorption of a certain amount of energy.

The instruments and operations of our universe are terribly complicated from the point of view of a postulational formulation. Such a formulation ideally should be in terms of the results of concrete individual instruments operated according to unique procedures. Actually we depart far from the ideal, for the "pressure" gauges, for example, may be of a great range of sizes and types of construction. The justification for using indiscriminately in our equations the indications of such a multitudinous family of operations can only be the experimental justification that such operations have been found in practice to give indistinguishable results. One of the most important considerations determining the acceptability of a physical concept is the extent of the range of operations, both instrumental and computational, which give indistinguishable results. The equivalence of different operations has to be subjected to renewed scrutiny as we progress toward the very small, and we have to be prepared in particular to find that different types of measurement, which under ordinary conditions may be equivalent, leave the "plateau" and enter the region of fluctuations at different stages. Thus an optical and a mechanical method may ordinarily be equivalent for determining the density of a colloidal suspension, but instrument makers found how to construct an optical instrument, the microscope, with which the fluctuations in the suspension could be seen at a much earlier stage in the development of physics than they could detect these fluctuations by "mechanical" methods. This difference in range of validity of different types of instrument I think is partly responsible for so much confusion about the significance of fluctuation phenomena.

Thermodynamics would hardly exist as a profitable discipline if it were not that the natural limit to the size of so many types of instrument which we can now make in the laboratory falls in the region in which the measurements are still smooth.

If one considers what is involved in a determination of the properties of the various parts of an unknown system, it appears that the universe of operations of conventional thermodynamics embraces pretty nearly all the operations which we can now perform, with the proviso that the instruments must not be too small and that the operations of atomic transmutation must be excluded. An interesting question is whether it is possible to find sub-groups of operations in this universe which shall themselves define causal systems. If one is willing to restrict the matter with which one is concerned to certain specific sorts, then such sub-groups are obviously possible, and these sub-groups correspond to the various recognized separate disciplines of physics. Thus the subject of mechanics utilizes a certain sub-group of operations. But the objects with which mechanics deals have to be all frictionless, and have perfect coefficients of restitution, for otherwise any actual system ceases to be mechanical because there is degradation of mechanical energy into heat. Similarly there is a sub-group of operations corresponding to optics, but if the system is to remain in the domain of optics it must be composed only of highly idealized perfect reflectors and non-absorbers. The existence of textbooks of mechanics or optics is sufficient evidence that it is profitable to make such idealizations and to break the universe of operations into sub-groups. But it is also evident, I think, that in nearly all actual cases the failure of actual materials to meet the idealized conditions is so great that the deviation from ideal properties is readily established with ordinary instruments, and the results of one's calculations have to be used with an instructed caution. It is only the universe of operations of thermodynamics which is so broad that its results can be used without onerous restrictions and qualifications. From this point of view the universe of operations of thermodynamics is itself a sub-group of all the operations which we can now perform, including operations of all scales of magnitude. This sub-group is sufficiently important to give thermodynamics its admitted great generality, but obviously in such a setting we are not to expect complete generality for it. Nevertheless, complete generality is postulated for the energy and entropy concepts, which are extrapolated from thermodynamics.

Another sort of extension of the universe of operations suggests itself. May there not be instruments and operations on a much larger scale than any that we can directly handle — instruments and operations such that it would be possible to describe new "causal" system in terms of them? Since any answer to this question must itself be "operational" it is pretty evident what the nature of any possible answer must be. Any such larger operations must be in the paper and pencil domain. It is quite conceivable that there should be paper and pencil combinations of results which we get with our present instruments which have simple and useful relations to other combinations of paper and pencil results. Of such a nature is an analysis of the gravitational properties of the universe using a mean figure for the number of stars per cubic parsec in the same way that the density of matter under ordinary conditions is assumed to be uniform. In fact, all cosmical calculations are almost necessarily of this character, and the extent to which they are successful is a measure of the degree to which it has been possible to find universes of larger operations defining causal systems. In the construction of these larger paper and pencil universes it may be necessary or convenient to postulate rules for the paper and pencil manipulations which would correspond to what would ordinarily be described as "emergent" properties. Thus, imagine that our instrumental skill had not yet reached sufficient perfection to permit direct detection of the gravitational attraction between any masses small enough to permit direct manipulation. Our calculations, however, would demand a "force" between bodies of celestial size. It might be convenient as a matter of calculation to postulate that this could be found in terms of action between small elements, but until the experiment had been made, we would not know whether this force was something that "emerged" only when we were concerned with masses above a certain size, or whether the assumption of a further significance was "real." Evidently the verbal element preponderates in this situation.

A question to be considered is whether there may not be new disciplines, with new sets of concepts, for universes of super-operations, corresponding to our classical thermodynamics for the universe of ordinary operations. The question will have to be answered by observation and by paper and pencil experimentation. A suggestion as to the possibilities is afforded by the new methods of treating turbulence in hydrodynamics. Turbulence is recognized as a parameter of a mass of liquid, which has its own rules of propagation and methods of characterization obtained by special combinations of the operations of thermodynamics, these latter analyzing the fluid into elements each with uniform physical properties and uniform velocity. A general requirement for such universes of super-operations would seem to be the existence of some corresponding sort of plateau phenomena.

There is one type of operation which obviously must be included in the complete universe of operations of thermodynamics, but which requires special treatment. We have seen that the concept of "recoverability" plays a fundamental role in thermodynamics; there must then be some way of knowing when an initial configuration is recovered, and this involves keeping somehow a record of the past. This record involves some sort of operation. But if the universe of operations includes the operations by which the past acquires meaning through some sort of record, then the universe can never be brought back completely to its initial configuration after any process, because the first time the process is performed the record must read that a certain configuration is now realized for the first time, and on the recovery the record reads that the configuration is now realized the second time. That is, when the record is made part of the universe no configuration can ever be recovered. The way out is, of course, in the assumption always made that the operations by which the records are kept, and in general everything dealing with the private life of the manipulator, plays no part in the processes of thermodynamics. This is doubtless legitimate enough under ordinary conditions, because the effects are so small, but I think it should be examined a little more carefully whether there may not be trouble here when we go to certain limits. This consideration certainly makes trouble if one seeks a logically rigorous set of postulates.

We have been concerned mostly with limitations of our instruments and operations imposed by size. Another type of limitation should obviously be considered, namely that imposed by time. It takes time for a thermometer to come to equilibrium with its surroundings and give the correct reading. How long shall we wait for the reading? Here again the only answer which experience has to give is a fuzzy answer. It is evident that we have the same "plateau" phenomenon with regard to time that we have with regard to size. At first the thermometer settles down to a smooth reading, but if we wait long enough and the thermometer is small enough, we shall presently get something different from the ordinary fluctuation effects, namely a kick due to the disintegration of some atom; or if the thermometer is larger and we wait very much longer we shall presently find the thermometer itself as well as its environment in the process of spontaneous transmutation. We assume without question that thermodynamics applies to the first plateau. We may if we like try the assumption that it also applies to any plateau there may be when the spontaneous processes of transmutation have run sufficiently to set up any characteristic equilibria they may have. Calculations have as a matter of fact been made based on such assumptions. Since the times involved in such equilibria are so terribly beyond the reach of direct experience, the only meaning we can attach to the "truth" of such an assumption is to be found in the indirect consequences drawn by pencil and paper operations from the assumption. It is probably too early to say whether thermodynamic methods may be applied in general to such long-term equilibria or not. From the paper and pencil point of view of statistics it is possible to regard all phenomena which we now describe as "mechanical" as special kinds of plateau phenomena in a setting which would eventually become "thermal" from the point of view of a time scale extravagantly longer than that available to us.

It is evident in the meantime that there are many plateau phenomena on a much shorter time scale, and that the assumption of the applicability of thermodynamics to these phenomena is made and is justified. For instance, consider the assumption of the existence of a membrane of zero thermal conductivity and zero thermal capacity which is used so freely in the arguments by which the Carnot cycle is defined and the absolute scale of temperature established. It is one of the best-established results of experience that there are no true thermal insulators and that any system comes to uniform temperature after the lapse of a sufficient time. How then can one be sure that the assumption of a perfect insulator does not violate some cardinal principle of thermodynamics, or have confidence in the results of making such an assumption? I think the answer is to be partly found in the existence of plateau phenomena.

THE ASSUMPTION OF MATTER WITH SPECIAL PROPERTIES

The same sort of question as considered in the last paragraph arises rather often, namely the question as to the justifiability of assuming in some thermodynamic argument the existence of matter with special properties. The situation is complex, and I do not believe that any one simple answer can be given. There are situations in which there is still genuine difference of opinion among competent physicists. Thus to many the permissibility of assuming the existence of a substance satisfying the equations of an ideal gas in order to prove the second law of thermodynamics and establish the absolute temperature scale, as was done by Planck and many others, still seems highly questionable. The Maxwell demon really amounts to the assumption of matter with special properties, an assumption which seems to me to be highly suspect. There are other situations in which at first the propriety of an assumption was seriously questioned but it is now accepted. Such an assumption was with regard to the existence of perfect semi-permeable membranes. The situation with regard to the semi-permeable membrane was, I think, never as bad as it appeared to the agitated defenders of thermodynamic rigor. The question was never the bald one of the existence of such membranes. All thermodynamics had to say was that *if* there were such membranes, then there must be another related physical phenomenon, osmotic pressure. There are other similar situations, in which it can be shown that if matter of certain novel properties exists then thermodynamics demands other related novelties. It then remains only to search experimentally for matter with any of these novel properties.

With regard to the permissibility of assuming perfect non-conductors I suppose that there is no difference of opinion. The argument involves a number of considerations, which might be put somewhat as follows. There are in nature materials of a wide range of thermal conductivities. Some of these are such poor conductors that the heat that would be actually conducted across the membrane in the part of the Carnot cycle when the insulation is supposed to be perfect is only a small part of that which passes in other parts of the cycle when the membrane is removed. Even with use of the same material it is possible to vary over a wide range the ratio of the unwanted to the wanted heat transfer by varying the dimensions of the apparatus. It is possible by special construction to get shields for thermal transfer of very small capacity, such as vacuum shields with silvered surfaces. No essential difference in the functioning of the engine seems to be detectible whether thermal transfer occurs by the mechanism of conduction or by the entirely different mechanism of radiation. No thermodynamic analysis has ever yielded any numerical values for a thermal conductivity, nor has it been able to establish relations between thermal conductivity and other quantities which do come within the scope of thermodynamic analysis. It would appear therefore that thermodynamics must still be applicable in the limiting case where the thermal conductivity vanishes. This perhaps is the favorite sort of argument, but it presupposes a considerable experience in thermodynamic analysis which cannot be assumed when one is entering fresh ground. The neatest way of dealing with the situation would be to write the complete equations for some actual physical set-up, in which occur among other things explicit symbols for the conductivity of the screens, and to show that the result which was obtained by the ordinary analysis in the special case where the conductivity is zero continues to hold no matter what the conductivity. In those cases where we have no question of the propriety of our assumptions I think that we almost always have an unexpressed feeling that this sort of thing must be possible if we bothered to carry it through. But in many cases the complications of such a complete analysis would be serious, and there are even many cases where we would be embarrassed as to how to proceed because the introduction of actual materials means the entrance of irreversible processes, as in the case of thermal conductivity, which traditional thermodynamics has not attempted to handle. In many cases difficulties of this sort would be avoided by the extension of the classical statement of the second law which has already been discussed, namely by postulating that there are certain essentially irreversible processes which are accompanied by their own characteristic increases of entropy wherever they occur. It will be found that the analysis of any system in which there is dissipation by thermal conduction can be made rigorous in this way.

In any event, whatever the nature of the special process assumed, it must not violate any of the broad generalizations of experience. Thus in establishing the absolute temperature scale it would not have been admissible to assume an engine reversible in the limit if the existence of such an engine could have been shown by *any* line of argument to involve an uncompensated increase of temperature differences in an isolated system. And since in general we can never know when we have followed through *every* line of argument, our feeling of security must be determined by the general consistency of our entire structure.

DETAILED BALANCING

A group of principles, somewhat similar to each other, has often been used in thermodynamic or statistical-kinetic discussions, which are called by various names by different authors, but may be broadly lumped together as principles of "detailed balancing." Thus, one broad form of the principle states that if some process is a complex process in the sense that different independent mechanisms can be recognized to be operative, then the final equilibrium attained in any actual system when the different mechanisms are operative jointly is the same as if any one of the mechanisms were alone concerned. Thus a collection of objects initially at different temperatures in an isolated enclosure comes to the same final temperature whether the method of heat transfer between the different members is conduction or radiation. A form of the principle is often used in theoretical analysis, as in Lorentz's famous discussion of an electron gas in which it was required that at a given temperature the same distribution of velocities should result whether the reaction between the electrons was collisional in nature or via the radiational electromagnetic field, or in Einstein's analysis of equilibrium in an atomic system in which transitions are occurring both by quantum jumps with radiation initiated by collisions and spontaneously.

It is important that we should be able to separate the instrumental from the paper and pencil in these situations. In order to be sure that what we are concerned with has thermodynamic significance, we should like to be able to define "independent mechanism" in terms of the macroscopic operations of thermodynamics. This is sometimes obviously possible, as when we speak of the attainment of thermal equilibrium by a mechanism of conduction or of radiation. It is true that even here there is no instrument that we can insert into the medium and which will indicate by the position of one pointer the amount of thermal energy being transferred by radiation and by another the transfer by conduction. If we could perform such a direct instrumental measurement we would have perfect confidence in talking about a "mechanism" or a "process." In most situations such a direct attack is not possible, but it is possible to proceed by more indirect means. In a system which is the seat of thermal transfer we can construct in imagination an infinitely finely divided cellular structure of infinitely thin perfectly conducting and perfectly reflecting membranes, and in this way eliminate transfer by radiation, leaving only transfer by conduction. In the laboratory we can approximate to this imagined state of affairs, and thus give a degree of instrumental meaning to the two independent mechanisms. The splitting into two mechanisms determined by such instrumental procedures is made more precise by the accompanying mathematical analysis. Thus the results of a family of experiments with cellular structures of different size might be covered by a single mathematical equation with terms whose coefficients vary in magnitude when the mesh varies. The terms in the equation to which the variable coefficients belong would then be the mathematical equivalent of the different "physical" mechanisms; mathematically, this is the meaning of "mechanism." A great advantage of the mathematical analysis is that it makes unnecessary variation of the physical conditions through the entire range from complete suppression of the one mechanism to complete suppression of the other, a condition almost impossible to realize, but it is necessary only that the contributions by the two mechanisms vary over a range. The essential requirement is always that variation be possible, either in the properties of the system or in the sort of instrumental operation to which the system may be subject. It is conceivable that a mere change of dimensions would suffice. The same sort of mathematical procedure makes it possible to dispense with other ideal conditions in the experimental set-up; no membrane has vanishingly small thermal capacity, or is a perfect reflector, or a perfect conductor. By using a number of membranes whose properties vary over a range the mathematical analysis may be made to give the effects of the different factors separately by an extrapolation. This sort of analysis is naturally most easily done when the relations are linear.

It is easy to see why in this example of thermal equilibrium the final equilibrium temperature must be independent of the mechanism. For although the equilibrium is attained by the detailed play of the mechanisms, the final state toward which this play is directed must satisfy certain broad thermodynamic conditions. Thus for a gas in an impervious enclosure the final temperature is completely and uniquely determined by the condition that the total energy of the system has a given value and that the pressure is uniform. Since these conditions which determine the temperature contain no mention of the mechanism, the temperature must be independent of the mechanism. The same result must hold in any other situation in which the final state is determined by broad thermodynamic or mechanical conditions which can be formulated without reference to any special mechanism. It is a rather special application of the principle of "detailed balancing" which is made to situations as general as these, and probably it would be better to use a different name for it.

Usually we are not content to say as little about the details of the mechanism as we did above. Thus we may say that the "mechanism" of heat conduction in a solid is partly by elastic waves. To what extent can the presence of these waves be shown by the macroscopic instruments of thermodynamics? It is of course easy with macroscopic instruments to show that a solid may transmit elastic waves, to measure the velocity of these waves, and to show that at least over a wide range this velocity is independent of the frequency. But if I apply the instruments with which I establish these results to the table top before me, it is dead. Instruments of microscopic, not macroscopic, size are needed to reveal the elastic waves of thermal conduction, for their wave lengths cluster around atomic magnitudes, and the energy of any such wave is $\kappa\tau$, which is the energy of fluctuation phenomena, and thus by definition beyond the reach of macroscopic instruments. Eventually we shall perhaps have a laboratory physics of microscopic instruments, but for the present the meaning of elastic waves of heat conduction is almost entirely a paper and pencil meaning, and the principle of detailed balancing applied to such waves is a principle that controls paper and pencil manipulations, in the realm of statistics and kinetic theory. What is the paper and pencil meaning of saying that heat is conducted in a solid by the mechanism of elastic waves? or in saying that there is also another mechanism, radiative in character? The meaning is to be found in the form of the equations with which we theorize about thermal conduction, and has already been suggested. Thus to assume an extreme simplification, we might find that the conductivity of all solids could be expressed as the sum of some function of its elastic constants plus some other function of its optical transparency. The checking of the equation demands variation in the properties of the solid, perhaps by making up various mixtures. It would then be the task of a theory to "explain" how the particular function of the elastic constants arises, and this explanation would amount to discovering the "mechanism" of conduction by elastic waves. The paper and pencil operations which give any mechanism of this sort meaning involve a background of theory, either macroscopic, or more usually, microscopic and kinetic.

The ideas of "mechanism" and of "independent mechanism" are very closely related. I think we would not say that a mechanism corresponded to a term in an equation unless physical systems were possible in which the contribution of this mechanism could be varied arbitrarily, keeping the contributions of any other mechanisms constant. On the other hand, if the equations contain groups of terms which always bear the same relation to each other under all physical conditions, then the analysis which suggested the use of several terms was a primarily verbal analysis, which we can presently learn to dispense with, and embrace the group of terms into a single concept with only one corresponding "mechanism."

Accepting now that there are two independent mechanisms in the paper and pencil sense which account for heat transfer in a solid, namely transfer by elastic waves and by radiation, then the principle of detailed balancing might make such a statement as this: "In a solid in thermal equilibrium the amount of heat transferred by elastic waves from right to left must exactly balance that transferred by elastic waves from left to right, and there must be a similar balance between the amounts transferred by radiation from right to left and from left to right. It is not possible that a greater elastic transfer from right to left be compensated by a greater radiative transfer from left to right." This is, either mechanism by itself must satisfy the requirement of thermal equilibrium that there be no net transfer of thermal energy in any direction. This is the sort of thing that is more usually understood by the principle of detailed balancing rather than the more general thing discussed above.

Why is it that there must be this exact balancing of each process independent of the other, and why is a properly controlled interaction between them impossible? If only a single physical set-up were possible, there seems no reason why an excess transfer by elastic waves in one direction should not be balanced by an excess radiative transfer in the other. The necessity for detailed balancing only arises when we have the possibility of different physical systems in which the magnitude of the transfer by elastic and radiative mechanisms varies independently. For suppose we have one physical system in which a net transfer to the left by elastic waves is compensated by a net transfer to the right by radiation. Then because of the possibility of independent variation another physical system is possible in which the transfer by elastic waves is the same but that by radiation is doubled. Such a system would obviously be out of equilibrium, which is contrary to experience. That is, the principle of detailed balancing must hold in situations like this because of the possibility of independent variation of the different mechanisms. But it is precisely this possibility of independent variation that gives meaning to the mechanisms themselves. It would thus appear that if it means anything to talk about mechanisms or processes, then the principle of detailed balancing must apply. The principle in the usual sense conceals a tautology.

The "principle of microscopic reversibility" as discussed by Tolman is somewhat different. This is almost exclusively a paper and pencil principle in the sense that the operations which give it meaning are operations in the domain of statistical analysis. The argument runs that any assemblage in statistical equilibrium contains as many members with any particular velocity as with the exact reverse. Since now the assemblage is a completely haphazard one there can be no correlation between the position and the velocity coördinates of the particles, so that another assemblage in which all the position coördinates are the same and the velocity coördinates exactly reversed would be indistinguishable statistically, and therefore indistinguishable in its macroscopic behavior. This means in particular that the future history in any specified macroscopic surroundings would be indistinguishable. The usefulness of such a principle can best be learned from a detailed examination of the use that is made of it in statistical calculations; this is not the place for such an examination.

One method of making a physical analysis into independent mechanisms is by lowering the temperature to near the absolute zero. Here there may be a great difference between the speed with which different parts of the system come to equilibrium. Thus recently a number of speculations have involved the fact that under laboratory conditions the effective temperature of the electrons may be very much higher than that of the atoms. In fact, the slowness of attainment of equilibrium by the electrons may be so extreme that a modification in the method of analysis becomes necessary. One may retain the thermodynamic parameters such as temperature and entropy in discussing the atomic parts of the system, but the electrons will have to be described with "mechanical" parameters, and their energy manifestations described under the heading of "work." This is an example of the limitations on the time of the operations permitted in the universe of operations, and the fact that "mechanical" phenomena may be merely plateau phenomena in a world which is "thermal" from the point of view of a much longer time scale.

PARADOXES OF THE ABSOLUTE ZERO

The effect of the time factor on our operations as we approach absolute zero demands more consideration than has usually been given. The rate of many natural processes is drastically changed; this is usually in the direction of a decrease, as for example in the attainment of radiative equilibrium, but there are cases where there may be an apparent increase due to a change in the physical constants of the materials, as for example the thermal conductivity becomes high and the heat capacity low. The problems raised by this change in velocity are by no means academic, as shown by the shifting equilibrium between para- and ortho-hydrogen or by the difficulty just mentioned of transferring heat between electrons and ordinary matter and so of attaining temperature equilibrium in the domain of magnetically low temperatures. This change in velocity may well demand a reëxamination of some of our accepted procedures. For instance, we have justified the assumption of thermal insulators at ordinary temperatures by a limiting process, starting with a conductor and in the limit letting its conductivity vanish. But suppose it is a question of a thermal insulator at absolute zero. We cannot start with our system at absolute zero, and then allow the conductivity to vanish in the limit. Absolute zero is unattainable, and conclusions as to what will happen "there" are merely conclusions about limiting behavior as we approach lower and lower temperatures. We are here concerned with two limiting processes, therefore, and the question of double limits arises. In what order shall we perform our two processes, and is the order immaterial?

What happens to friction at zero absolute? One may

get into trouble in his thinking about it if he tries first to imagine a body already at absolute zero, and then inquires what happens to the mechanism of friction. For friction involves a dissipation of the elastic waves, which have their origin in the slipping past each other of the irregularities in two opposing surfaces, into the irregular waves of heat. The dissipative mechanism is provided by the irregular motion of the atoms due to temperature agitation. There is no dissipation at absolute zero therefore, and friction must become a phenomenon of mechanics. This should mean that the phenomena of friction become reversible. It will still take a force to drag one surface over another when there is a normal pressure between them, so that the phenomena can still be expressed in terms of a coefficient of friction. But one body having been dragged to the left over another by the exercise of a tangential force of five pounds, let us say, will spontaneously retrace its path back to its original position if the force is decreased infinitesimally below five pounds. The reason is that all the elastic waves which are generated during the forward motion by the pushing aside of the surface irregularities retain their form unaltered ready to push back when the motion is reversed. The world at absolute zero thus presents some highly paradoxical aspects; the position of objects on a level table top could not be changed by sliding them about, but the moment we took our hands off they would slide back to their original positions. If we approach close enough to absolute zero we may perhaps expect to find the first beginnings of this paradoxical behavior. The manifestation will be in transient phenomena which occur in the time interval during which the dissipation of the elastic waves is not complete. Ordinarily the time of manipulation of macroscopic objects is very much greater than the time of dissipation of elastic waves of the scale of magnitude of the surface irregularities. Here is a realm of a new order of phenomena at temperatures very much below those at present accessible through magnetic phenomena. There will be one thermodynamics for operations so rapid that tangential displacements are sensibly reversible, and another thermodynamics for operations so slow that frictional dissipation is sensibly complete. The approximate attainment of complete reversibility in frictional phenomena at absolute zero will doubtless be prevented by the zero-point energy and the fuzziness in position and velocity demanded by the Heisenberg principle.

There is a paradox of the absolute zero connected with the operation of reversible heat engines. We have the relation between the heat absorbed by the engine from the hot reservoir, the temperature of the hot reservoir, the heat discharged by the engine to the sink and the temperature of the sink: $Q_1/\tau_1 = Q_2/\tau_2$. If now the temperature of the sink is made zero absolute the heat discharged to the sink becomes zero also, and all the heat taken in by the engine from the reservoir is converted into work. Furthermore, since no heat is delivered to the sink, the sink does not increase in temperature as the engine continues to run, unlike the situation at any temperature above zero. That is, apparently we have here a state of affairs that is capable of indefinite self-maintenance, and there is an obvious violation of the intent of the formulators of the second law of thermodynamics. One suspects trouble with the order of his limiting processes, for the size of the reservoir appears to make no difference. So long only as the reservoir is at absolute zero it receives no heat from the engine and cannot therefore increase in temperature, no matter how large the engine or how fast it works, or whether the sink is no larger than a pea. Why not, therefore, make the sink indefinitely small, and in the end dispense with it altogether?

There is, of course, no trouble as the situation presents itself in the laboratory. Bodies at absolute zero or even very close to zero which we can use as sinks do not occur in nature. If we want to operate an engine with a sink at a very low temperature we first have to make the sink. If we work through the details we find that we always have to do as much work to pump the heat out of the sink and get it ready for use as we gain afterward from the engine. The degree to which we can work the engine afterwards always is determined by the size of the sink which we have previously got ready. This holds exactly at all stages of the limiting process as we approach zero, and physically must hold always. But in our argument we imagine ourselves already at zero degrees: we said "at zero degrees no heat at all (or zero heat) is given to the sink, so that its size is immaterial, and in particular it may be of infinitesimal or zero size." The logic of this argument is a spurious verbal logic, derived from the implications of "zero" in other contexts.

Another difficulty at absolute zero might at first appear insuperable, namely the difficulty of adequately describing any system. Ordinarily our information is obtained by optical methods, and at zero the density of black-body radiation vanishes. There is, however, a simple solution, namely to obtain our information about a system by monochromatic radiation. Manipulation of such radiation is to be classed as a mechanical operation, and there is no limitation at any temperature on the density of such radiation. In fact, at all temperatures, the obtaining of information about a system should be called a mechanical operation.

BIOLOGICAL PHENOMENA AND THERMODYNAMICS

The view that the universe is running down into a condition where its entropy and the amount of disorder are as great as possible has had a profound effect on the views of many biologists on the nature of biological phenomena. It springs to the eye that the tendency of living organisms is to organize their surroundings, that is, to produce "order" where formerly there was disorder. Life then appears in some way to oppose the otherwise universal drive to disorder. What is the significance of this? Does it mean that living organisms do or may violate the second law of thermodynamics?

There has been an enormous amount of fuzzy thinking on this subject, much of which has resulted from a misconception of what thermodynamics says. The more mystically minded like to find the significance in the "volitional" properties of living things, which of course almost inevitably leads presently to talk about "free will." The impulse to this unfortunate sort of loose speculation might well have been initiated by Maxwell's demon itself, for it is easy to say that the ability of the demon to violate the second law came essentially from his ability to open or close his trap door by an act of volition in accord with the velocity of the approaching molecules. But this is after all a superficial point, because we could certainly construct a robot who would automatically open the door at the right time. That any hypothetical ability of living beings to violate the second law cannot reside in any supposed power of volition as such is apparent enough when I consider that I, a being with "volition," do not know how to set to work to violate the second law even though I may allow myself any arbitrariness of construction and action of which I may be capable. If I am violating the second law I do it unknowingly and without the capacity for taking advantage of it. The more one thinks of it, the more nebulous and purely verbal appears this thing called volition ascribed to living things. I can see no way not trivial of giving operational meaning to "volition." In any situation which repeats a previous situation a robot may always be constructed — Newcomen's pumping engine which originally operated with a small boy to pull open the valve at the end of each stroke evolved into the completely automatic engines we know.

Many of the set-ups proposed for exhibiting the relation of living things to the second law do not properly reproduce the conditions necessary for the application of the law. For instance, the environment of most living things is a stream of radiation from the sun to the earth, from which they extract energy which is used in the "organization" of the environment. The stream itself is a factor with "order" in the determining conditions; to prove that the second law has been violated would demand a quantitative proof that the "order" created by the organism in the final product is greater than the order in the stream of energy which made the process possible. A rigorous set-up for testing the second law might run somewhat as follows. Given, an isolated chamber which with its contents has come to equilibrium at a constant temperature. Suppose now that a minute spore is introduced into the chamber or is generated spontaneously. The spore starts to grow and creates an organism of some size by synthesis from its surroundings both material and radiative. The organism thus generated is negatively geotropic; it climbs to the top of the chamber against gravity, and there completes its life cycle, dies, and decomposes to the elements from which it came, leaving its dead remains at the level to which it had climbed. The net result as far as the chamber is concerned is that it is now capable of delivering mechanical work by the falling of the remains of the animal. This is contrary to the second law, because the energy delivered by the falling of the weight has come from the miscellaneous temperature energy of a system at uniform temperature.

As far as I know no attempt has ever been made to carry out any such cycle; certainly if any successful attempt has been made to examine on rigorous grounds the applicability of the second law to biological systems the result is not generally known. Any direct experimental attacks on this problem appear to be very meager. It would be possible to devise experiments which would be instructive, although much less far-reaching than the one suggested. One of the tasks of elementary thermodynamics is to derive certain relations between the various properties of bodies. Some of these relations involve in their deduction only the first law of thermodynamics, while others involve the second law. An interesting experiment would be to test whether the relations which involve the second law are satisfied by living matter. Consider, for example, an electric cell delivering a current. When current flows the cell tends to become warmer or cooler (whether warmer or cooler depends on the particular kind of cell), so that heat has to be put in from outside or else abstracted if the temperature of the cell is to remain constant while it is running. Now it is possible to establish with the help of the second law that there is a simple connection between the heat which must be given to the cell to keep its temperature constant and the change of the electromotive force of the cell when the temperature of the whole cell is raised. An exact parallel as far as the analysis goes can be set up between the electric cell and a biological organism delivering energy at the expense of the energy of the materials on which it feeds. An immediate check of one of the consequences of the second law could be made on the organism if it were not for the complicating fact that the organism itself undergoes changes while it is delivering energy which it would be difficult to evaluate in terms of energy or entropy. If we could find some living thing that would not itself change, or would change in a measurable way as does the battery, we could make an immediate check. Most organisms are hopelessly complicated, but the biologists have suggested a possibility to me. The excised heart of a turtle lives for a long time, continually beating and delivering energy as long as it is maintained in the proper nutrient medium. The rate of beating of such a heart depends on its temperature, as the electromotive force of a battery depends on the temperature of the battery. When such a heart delivers work by beating it should tend either to warm or cool, as does the battery, and it should be possible to measure the connection between the heating effect and the temperature coefficient of activity. In this way it might be shown that the second law is so far applicable to living things.

Although the experiment has not been made, I think that most biologists who have thought much about the problem would be surprised if the check were not forthcoming. I think they would first look for any possible violation of the second law in the realm of small things rather than in the macroscopic domain as just suggested. The question is: are living things capable of sorting operations, perhaps on the pattern of the Maxwell demon, which could perhaps be made to result eventually in such large-scale situations that useful work can be gained? There is evidence that organisms are capable of a kind of sorting process as an incident to the passage of food through their systems; the cycle of assimilation and elimination does not always return the material to exactly its initial configuration. Such for instance is the splitting of racemic compounds into right- and left-handed forms. But before this can be worked into a proof of a violation of the second law some method must be devised for extracting work out of the conversion of right- and left-handed forms back to the racemic form, and then it must be shown that the work so gained is greater than that lost in the processes necessary to the life and functioning of the organism. We appear to be far from the quantitative command of the situation that would make such a proof possible.

Now that we have quantum phenomena a proof that vital phenomena are outside thermodynamics would not be so catastrophic as it would have been earlier in scientific history. All it would mean would be that the operation of planting a yeast spore in a culture medium and letting nature take its course is not one of the universe of operations of thermodynamics. But as far as I am aware there is not at present the slightest experimental indication that such a conclusion will ever be necessary, or any reason to think that vital phenomena do not run their course on a level above the atomic and molecular. The impulse to think the contrary, so far as its origin is scientific and not mystical and animistic, is in a vagueness of apprehension of the conditions under which the second law applies and a confusion of the "order" of thermodynamics with the "order" of daily life; the latter has implications of esthetic creation and of "purpose."

If the problem of the artificial creation of life should ever be solved we would be far on the road toward knowing definitely whether organisms obey the second law or not. If we could assign a definite entropy to an organism, we could at once answer our question about the second law. To assign an entropy to an object demands some reversible method of getting to the object from a standard starting point, and this, for an organism, is close to the problem of the artificial creation of life. My own feeling is that any entropy which we may eventually have to assign, in virtue of its life, to an aggregation of elements into an organism will be only a small part of the entropy which we would now assign from the point of view of present macroscopic operations to approximately the same organism dead.

A new idea has been interjected by Bohr into our conception of the possibilities in the way of treating vital phenomena scientifically, that is, reducing vital systems to completely causal systems. This idea is that there may be a vital analogue for the Heisenberg principle. Specifically, it may be that any physical instrument or any physical process which we may be able to use as a probe in our search for an adequate description of what constitutes life must necessarily destroy the life that we are attempting to characterize, so that we are doomed to a perpetual ignorance. Or, put in another way, this would simply be saying that the question "what is the explanation of life?" is meaningless. It is easy to see the attractiveness of a suggestion like this; it rounds out the picture and presents a symmetry between living and non-living phenomena which is esthetically pleasing and which offers possibilities for further intellectual exploration. But it seems to me that the positive impulse to accept such a picture comes almost entirely from the formal intellectual side. Whether we think it probable that any such limitation will be eventually established in the laboratory and build our programs of speculation and experimentation on such an expectation must in the present state of biological knowledge be almost entirely a matter of personal taste. Personally I see such enormous complexities possible in the organization of matter above the atomic scale that I believe that a program which envisages nothing fundamentally different from what we now know in "physical" and "chemical" phenomena can be profitably followed for a long time.

In the introduction I have already commented that thermodynamics smells more of its human origin than other branches of physics — the manipulator is usually present in the argument, as in the conventional formulations of the first and second laws in terms of what a manipulator can or cannot do. Or, if the manipulator is not obviously present, he is negatively present in the statement that this or that process occurs "spontaneously," for what is "to occur spontaneously" except to occur without the connivance of a manipulator? The tool of the manipulator is often the "machine" or "engine," surely something which is not freely presented to us by a bounteous nature. In fact, the ubiquity of the manipulator in thermodynamics might tempt us to think that we have here an anticipation of the discovery by wave mechanics of the importance of the role of the observer and of the act of observation. I believe, however, that an examination of the details of the way in which the manipulator or the machine enters the arguments of thermodynamics will show that the spirit of the whole enterprise is fundamentally different from that in wave mechanics. There is here no opposition or contrast between what may happen spontaneously in nature and what may be brought about by the manipulator with his machine, but rather there is the most intimate assimilation between inanimate nature and the animate manipulator — no special characteristics are imputed to the manipulator because he is alive. If the statement is made that certain things do not occur in nature, then it follows that no manipulator with any sort of machine can make them occur. And conversely, any sort of thing that a manipulator can make occur with the help of any sort of machine may be expected to be found sometime freely occurring in a primitive nature. The attitude of the good bishop Paley is reversed: if I know that watches can be made by men in shops, then I must be prepared to find nature putting together a watch sometime on her own in the desert. The use of the manipulator in thermodynamics is merely a psychological device for getting me to think of all the things that might possibly happen.

Our first impulse in the paragraph above was to identify the "spontaneously" of thermodynamics with detachment from a manipulator; now that we have deanimated the manipulator we must deanimate the "spontaneous" also. Examination of the way it is used will show I think that to say that something happens spontaneously is merely another way of saying that it happens in an enclosure impervious to the passage of energy in any form. Under these conditions the manipulator has no special powers on which an untutored nature may not stumble by herself. Thermodynamics recognizes no special role of the biological.

ON "PHYSICAL REALITY"

The expression "physical reality" has occurred rather frequently in our exposition. The frequency with which I have wanted to use this expression has been rather a surprise to me because in my previous writings I have taken the position that the concept of "reality" is one that we must learn to get along without, and that the strength of one's impulse to use it is a measure of one's operationally weak-mindedness. My excuse for using the expression so often here, even though modified by quotation marks, is that the interest of the exposition has been partly in showing how verbal impulses determine what we actually do in physics, and I believe that as a matter of fact the impulse to use a "physical reality" plays a prominent part in the thinking and verbalizing of physicists. Although the words smack of metaphysics, I believe that an examination of the way in which the physicist uses them will show that he imposes fairly definite physical criteria, so that actually the metaphysical taint is slight. Because of the unfortunate metaphysical implications, however, I still think it would be best to avoid the word when possible.

We have talked of the "physical reality" of energy when localized at the points of a body or in a field, or the "physical reality" of a flux of heat or mechanical energy through a surface arbitrarily oriented at any point, or of the "physical reality" of the entropy which we associate with the points of a body or of the flux of entropy through a body. The common feature in all these things to which we ascribe "physical reality" is that they may be determined in terms of instrumental operations made at the point and the instant in question. A magnetic field, for instance, that could be obtained only by the integration of an expression involving the time derivative of the electric vector distributed through space would not be ascribed "physical reality." The magnetic vector acquires this status only when an independent instrumental procedure is found for getting it at the point in question. The "electrotonic" state of Faraday never acquired the status of "physical reality." As far as the requirement that the operations must be performed now goes, this means that such a thing as the "heat content" of a body has no "physical reality" as it was once thought to have. For although "heat content" may be given a definite meaning in terms of the time integral of the heat which a body has as a matter of fact received in its past history, such an integral has no unique connection with operations in which the body can take part now.

The rules by which the results of the instrumental operations are combined with paper and pencil may have any degree of complication so long as the rules are assignable in advance, and the instrumental operations themselves may have any degree of complication so long as they are local in space and time. When I started writing this essay, which has been revised a number of times, I thought that there were other requirements which we tacitly impose for "physical reality." For instance, when I observed that there is no one physical instrument or operation which will give the energy of a body, I thought I had thereby divorced energy from "physical reality." I now see that the usage of physicists is not as narrow as this; it is immaterial that the instrumental operations by which the kinetic energy of a body is determined are entirely different from those by which its thermal or electromagnetic energy is determined, or that the operations for assigning thermodynamic energy require preliminary rehearsal. This observation, by the way, is rather inimical to the old metaphysical idea that the "dimensions" of a physical quantity are an expression of its "essential physical nature." The use of a single word "energy" to cover such an operational diversity is justified because of the way in which the energy so diversely determined is used in the paper and pencil calculations.

Although physicists apparently do not place the same tacit limitations on the implications of "physical reality" that I at one time supposed, there nevertheless do seem to be some limitations. Thus I think any physicist would be unwilling to formulate any "general law" of physics in terms to which he could not ascribe "physical reality." One can see this by recalling how, in our discussion of the first and second laws of thermodynamics, we were continually at pains to assure ourselves that the terms of our formulations had "physical reality." There are other more vague and special requirements; thus we have seen that a "velocity" probably would not be imputed "physical reality," even although it might be uniquely determined by operations here and now, unless it transformed with the motion of the frame of reference in the same way that the velocity of a material particle transforms.

It is possible, I think, to see the common element that runs through these usages by the physicist of the concept of "physical reality," at least as far as the major requirement goes, and to understand to a certain extent why he finds the concept profitable. Probably the fundamental physical or mental operation is the operation of isolation. The first lesson the child must learn in finding how to meet his environment is to analyze out of it objects which have permanency and are independent of other objects, and similarly in learning the use of language he recognizes and isolates mental operations and situations for which he has words. Now the use of operations performed here and now is simply a special kind of isolation. The demand of the physicists for "physical reality" is pretty much the same as the demand for isolatability, and his meaning for "physical reality" is pretty much the same as "isolatability." If we want to look further, we may say that the strength of his insistence and his need for isolatability is one aspect of his universal need for simplification. It is, however, not a matter of intellectual *necessity* that I have isolatability and "physical reality." If I did not have them, I think I would not be stumped, but could make shift to get along somehow. But I do not see what I would do if I did not have some sort of *correlation*: whenever a present situation and the whole past that led up to it repeats I want to have at least some aspects of the future repeat. We could face a world as complicated as this without despair — perhaps we should make more provision for such a possibility in our thinking.

This I think is the inwardness of the sort of "physical reality" that we have allowed the physicist to talk about thus far. This view of the nature of the concept is consistent with the observation that in practice it is not always sharp, although a naïve first impulse would be to say that a thing either is real or is not. Thus I think we are rather likely to ascribe "less" "physical reality" to something in which the connection of the final result to the instrumental operations is through a long and complicated series of paper and pencil operations than when the connection is more immediate. I think most of us would have the verbal impulse to ascribe a greater "physical reality" to the mass of a gram of water than to its absolute entropy as obtained by a statistical calculation by the Boltzmann formula.

In spite of the universality of the operation of isolation, it is not one which conventional mathematics finds congenial. To treat isolation mathematically would demand the use of discrete discontinuous functions, instead of the ordinary analytical functions. When we try even to describe a single concrete event by continuous functions, as in a development in Fourier's series, we are forced to admit a fringe rippling out through infinite space, or quivering premonitions and remembrances stirring through all past and future time. It is possible, by the way, to erect an epistemology on the Fourier series just as on the probability use of the ensemble.

"Physical reality" as used thus far is a fairly definite concept, undeniably used by physicists, and with a certain utility. Observation shows, however, that many physicists do not restrict their usage to these fairly well defined conditions, but extend it to include implications which seem to me little removed from pure metaphysics. There can be no question, I think, that a concept covered by the same name "physical reality" is back of the attitude of many physicists toward general relativity theory, and is responsible for their almost religious conviction of the naturalness and necessity of the theory. This, however, is not the place for a discussion of the background of relativity theory, which I have tried to give elsewhere.

CHAPTER IV

RETROSPECT AND PROSPECT

IF WE ATTEMPT to define thermodynamics through its subject matter, then it embraces a study of all phenomena to which the first and second laws apply. Since the first law assumes a knowledge of the changes of mechanical energy accompanying any change in the parameters of a system, thermodynamics properly embraces many branches of physics which are usually treated separately and in which thermal phenomena as such are not prominent, such as mechanics and electrodynamics, for instance. If we examine the conditions which give meaning to the concepts of the two laws of thermodynamics — energy, entropy, and temperature — then a more specific definition would characterize thermodynamics through its universe of operations. The instrumental or laboratory operations included in this universe are those which would ordinarily be described as macroscopic operations. There is no sharp limit as we proceed in the direction of the very small to what we mean by macroscopic operations, nor are the readings which we obtain with such instruments clean-cut and unique. In general, as we pass toward the very small, we encounter plateau phenomena; the readings smooth out and appear to approach a clean limit without fluctuations. In a very few cases we can penetrate beyond the plateau phenomena with physical instruments and find fluctuations beyond. Macroscopic instruments and operations are by definition those which do not get beyond the plateau. The physical indefiniteness concealed in this definition is in practice of little moment because of the very small number of physical operations which even now are capable of getting beyond the plateau; when thermodynamics was first formulated there were no such operations. Any physical indefiniteness does not get into the paper and pencil operations because the first preliminary to the paper and pencil operations is to replace the instrumental indications by numbers mathematically sharp.

Nearly all our physical experience comes within the scope of macroscopic attack. Even heterogeneous systems may be broken up into elements so small that they may be treated as homogeneous by instruments small by the standards of the kitchen but still macroscopic because they are not beyond the plateau. This great macroscopic field is the universe of thermodynamics; there are two broad generalizations which apply to what happens in it, the first and second laws. There is no way, however, of being sure that we have stated the generalizations in the best way or that there may not be others which we have overlooked. Thermodynamics deals with necessary as distinguished from sufficient conditions. It leaves untouched broad classes of "irreversible" phenomena which can be adequately characterized by macroscopic operations. One would expect that there should be some macroscopic method of connecting these with the conventional macroscopic subject matter of thermodynamics, and an extension of classical thermodynamics has been suggested which allows this.

It is not the function of thermodynamics to "explain" macroscopic phenomena; it is not concerned with the origin of the equation of state of a gas, but treats it as given.

"Explanation" appears to demand an excursion into the atomic domain; it operates with the concepts of statistical mechanics and kinetic theory and is to a large extent a paper and pencil affair. Controlling the operations of paper and pencil are certain rules extrapolated into the atomic domain from the macroscopic world of thermodynamics, in particular the conservation of energy. Practically every check which anyone has attempted to give of the legitimacy of this extrapolation is a macroscopic check; Avagadro's number is determined by operations with balances and meter sticks; the Stern-Gerlach experiment gives its results through a macroscopic optical operation on a photographic plate, and the single atomic disintegration of a Geiger counter is a crack of sound to my ears in a loud-speaker. The question of the legitimacy of the extrapolation *in the atomic world itself* is hardly touched by experiment. The whole enterprise of this extrapolation seems to be concerned not with the formulation of a method for getting into a new world but rather with the formulation of rules for getting back into the world from which we started. It is surprising that physicists are so satisfied with an "explanation" when the "truth" of the assumptions on which the explanation rests can hardly be given meaning. The situation is, I think, even less satisfactory with regard to the probability "explanations" of the second law. There is a logical chasm between the physical situation and the paper and pencil operations of probability theory. The probability treatment of the second law is useful and fruitful because the operations can be performed uniquely by us; such a treatment is really an exploitation of a universal feature of human psychology rather than an explanation.

These difficulties are not pressing at present, and we can

usually afford to ignore them because so very few of our present instrumental operations are in the region beyond the plateau. Such of them as are beyond the plateau give only very fuzzy results. It is encouraging that none of the fuzzy results that we now have appear to be inconsistent with our paper and pencil extrapolations. One may perhaps anticipate that we shall presently learn how to increase the number of instrumental operations possible beyond the plateau and sharpen their indications. A whole new world may be waiting here, whose properties it would be futile to attempt to anticipate. Perhaps biological phenomena reside in this world. It is hard to imagine how one would go to work to make a laboratory attack on this world. Probably we will have to use the fact that different classes of physical phenomena drop over the edge of the plateau at different stages; optical phenomena leave the plateau earlier than the phenomena of a microscopic dissecting needle. Perhaps the best way of starting an experimental attack would be an exhaustive study of all that we can do with dissecting needles under the microscope.

The possibilities left open to us by our present macroscopic knowledge may be visualized by an analogy. It is as if our entire present world were like what we see with our unaided eyes on the screen of a movie stage. This is an orderly world; events are causally connected and sweeping generalizations can be made about what happens, such as that stones do not fall upward. But suppose that we acquire command of new instrumental operations other than those we can perform with our eyes; suppose we can analyze the picture on the screen with a camera making a thousand exposures a second, or suppose that we can leave our seats and inspect the mechanism of the projecting booth. Everything is now altered; phenomena now appear of which we had no conception and which we would not have known how to formulate: the stationary frames, the phenomena of shift. New parameters appear: the number of frames per second, the velocity with which the film moves while one frame is replacing the next. By manipulation of the projector we would be able to nullify some of our sweeping generalizations, such as that stones always fall downward. And we would be able to explain some of the rare events in our naïve movie world which defied our sweeping generalizations, such as that sometimes wheels turned the wrong way. This latter is of course the analogue of our present fluctuation phenomena.

The analogy is not to be taken too seriously, but I think it has something in it. The stationary frames correspond to the fact that all we can get with our macroscopic instruments or treat with thermodynamics are the positions of approximate rest. The world of intermediate phenomena, of the possibility of whose existence we get some inkling through fluctuation effects, eludes us. No paper and pencil manipulations can introduce us to this world or even establish its existence or non-existence; for this we must, as always, wait for fresh developments in the laboratory. I think that about all we can claim at present is that we have no present knowledge that makes the existence of such phenomena impossible, and that if they do exist, we can see the place where they might be.

INDEX